Renewable Power Pathways

A REVIEW OF THE U.S. DEPARTMENT OF ENERGY'S RENEWABLE ENERGY PROGRAMS

Committee on Programmatic Review of the U.S. Department of Energy's
Office of Power Technologies
Board on Energy and Environmental Systems
Commission on Engineering and Technical Systems
National Research Council

NATIONAL ACADEMY PRESS
Washington, D.C.

National Academy Press • 2101 Constitution Avenue, N.W. • Washington, D.C. 20418

NOTICE: The project that is the subject of this report was approved by the Governing Board of the National Research Council, whose members are drawn from the councils of the National Academy of Sciences, the National Academy of Engineering, and the Institute of Medicine. The members of the committee responsible for the report were chosen for their special competences and with regard for appropriate balance.

This report has been reviewed by a group other than the authors according to procedures approved by a Report Review Committee consisting of members of the National Academy of Sciences, the National Academy of Engineering, and the Institute of Medicine.

This report and the study on which it is based were supported by Contract No. DE-FG01-98EE35047 (Task #2) from the U.S. Department of Energy. Any opinions, findings, conclusions, or recommendations expressed in this publication are those of the author(s) and do not necessarily reflect the view of the organizations or agencies that provided support for the project.

International Standard Book Number: 0-309-06980-7

Available in limited supply from:
Board on Energy and Environmental Systems
National Research Council
2101 Constitution Avenue, N.W.
HA-270
Washington, DC 20418
202-334-3344

Additional copies available for sale from:
National Academy Press
2101 Constitution Avenue, N.W.
Box 285
Washington, DC 20055
800-624-6242 or 202-334-3313 (in the Washington metropolitan area)
http://www.nap.edu

Copyright 2000 by the National Academy of Sciences. All rights reserved.
Printed in the United States of America.

THE NATIONAL ACADEMIES

National Academy of Sciences
National Academy of Engineering
Institute of Medicine
National Research Council

The **National Academy of Sciences** is a private, nonprofit, self-perpetuating society of distinguished scholars engaged in scientific and engineering research, dedicated to the furtherance of science and technology and to their use for the general welfare. Upon the authority of the charter granted to it by the Congress in 1863, the Academy has a mandate that requires it to advise the federal government on scientific and technical matters. Dr. Bruce M. Alberts is president of the National Academy of Sciences.

The **National Academy of Engineering** was established in 1964, under the charter of the National Academy of Sciences, as a parallel organization of outstanding engineers. It is autonomous in its administration and in the selection of its members, sharing with the National Academy of Sciences the responsibility for advising the federal government. The National Academy of Engineering also sponsors engineering programs aimed at meeting national needs, encourages education and research, and recognizes the superior achievements of engineers. Dr. William A. Wulf is president of the National Academy of Engineering.

The **Institute of Medicine** was established in 1970 by the National Academy of Sciences to secure the services of eminent members of appropriate professions in the examination of policy matters pertaining to the health of the public. The Institute acts under the responsibility given to the National Academy of Sciences by its congressional charter to be an adviser to the federal government and, upon its own initiative, to identify issues of medical care, research, and education. Dr. Kenneth I. Shine is president of the Institute of Medicine.

The **National Research Council** was organized by the National Academy of Sciences in 1916 to associate the broad community of science and technology with the Academy's purposes of furthering knowledge and advising the federal government. Functioning in accordance with general policies determined by the Academy, the Council has become the principal operating agency of both the National Academy of Sciences and the National Academy of Engineering in providing services to the government, the public, and the scientific and engineering communities. The Council is administered jointly by both Academies and the Institute of Medicine. Dr. Bruce M. Alberts and Dr. William A. Wulf are chairman and vice chairman, respectively, of the National Research Council.

COMMITTEE ON PROGRAMMATIC REVIEW OF THE U.S. DEPARTMENT OF ENERGY'S OFFICE OF POWER TECHNOLOGIES

H.M. HUBBARD (chair), Pacific International Center for High Technology Research (retired), Golden, Colorado
R. BRENT ALDERFER, consultant, Doylestown, Pennsylvania
DAN E. ARVIZU, CH2M Hill, Greenwood Village, Colorado
EVERETT H. BECKNER, Lockheed Martin Corporation, Bethesda, Maryland *(until December 2, 1999)*
PETER BLAIR, Sigma Xi, Research Triangle Park, North Carolina
CHARLES GOODMAN, Southern Company Generation, Birmingham, Alabama
NATHANAEL GREENE, Natural Resources Defense Council, New York, New York
JEFFREY M. PETERSON, New York State Energy Research and Development Authority, Albany
RICHARD E. SCHULER, Cornell University, Ithaca, New York
T.W. FRASER RUSSELL, NAE,[1] University of Delaware, Newark
JEFFERSON W. TESTER, Massachusetts Institute of Technology, Cambridge

Liaisons from the Board on Energy and Environmental Systems

ROBERT W. SHAW, JR., Aretê Corporation, Center Harbor, New Hampshire
JACK WHITE, The Winslow Group, LLC, Vienna, Virginia

Project Staff

RICHARD CAMPBELL, program officer and study director
JAMES ZUCCHETTO, board director
CRISTELLEN BANKS, project assistant *(until October 31, 1999)*

---

[1] National Academy of Engineering.

BOARD ON ENERGY AND ENVIRONMENTAL SYSTEMS

ROBERT L. HIRSCH (chair), Advanced Power Technologies, Inc., Washington, D.C.
RICHARD E. BALZHISER, NAE,[1] Electric Power Research Institute, Inc. (retired), Menlo Park, California
EVERETT H. BECKNER, Lockheed Martin Corporation, Albuquerque, New Mexico *(until December 2, 1999)*
WILLIAM L. FISHER, NAE, University of Texas, Austin
CHRISTOPHER FLAVIN, Worldwatch Institute, Washington, D.C.
WILLIAM FULKERSON, Oak Ridge National Laboratory (retired) and University of Tennessee, Knoxville
EDWIN E. KINTNER, NAE, GPU Nuclear Corporation (retired), Norwich, Vermont
GERALD L. KULCINSKI, NAE, University of Wisconsin, Madison
EDWARD S. RUBIN, Carnegie Mellon University, Pittsburgh, Pennsylvania
JACK SIEGEL, Energy Resources International, Inc., Washington, D.C.
ROBERT W. SHAW, JR., Arete Corporation, Center Harbor, New Hampshire
ROBERT SOCOLOW, Princeton University, Princeton, New Jersey
K. ANNE STREET, consultant, Alexandria, Virginia
KATHLEEN C. TAYLOR, NAE, General Motors Corporation, Warren, Michigan
JACK WHITE, The Winslow Group, LLC, Fairfax, Virginia
JOHN J. WISE, NAE, Mobil Research and Development Company (retired), Princeton, New Jersey

Liaisons from the Commission on Engineering and Technical Systems

RUTH M. DAVIS, NAE, Pymatuning Group, Inc., Alexandria, Virginia
GAIL DE PLANQUE, NAE, consultant, Potomac, Maryland
LAWRENCE T. PAPAY, NAE, Bechtel Technology and Consulting, San Francisco, California

Staff

JAMES ZUCCHETTO, director
RICHARD CAMPBELL, program officer
SUSANNA CLARENDON, financial associate
CRISTELLYN BANKS, project assistant *(until October 31, 1999)*

---

[1] National Academy of Engineering.

Acknowledgments

The Committee on Programmatic Review of the U.S. Department of Energy's (DOE's) Office of Power Technologies wishes to thank the many individuals who contributed significantly to this National Research Council (NRC) study. The presentations by representatives of DOE and the national laboratories, as well as a host of other organizations, provided the committee with valuable information and insights on DOE's Office of Power Technologies programs and the renewable energy technologies under development. These participants also participated in valuable discussions during question and answer sessions with the committee (see Appendix B for a list of presentations).

The chairman also wishes to recognize the committee members and the staff of the NRC's Board on Energy and Environmental Systems for their hard work in organizing and planning committee meetings and their individual efforts in gathering information and writing sections of the report.

The report has been reviewed by individuals chosen for their diverse perspectives and technical expertise, in accordance with procedures approved by the NRC Report Review Committee. The purpose of this independent review is to provide candid and critical comments that will assist the authors and the NRC in making the published report as sound as possible and to ensure that the report meets institutional standards of objectivity, evidence, and responsiveness to the study charge. The content of the review comments and draft manuscript remains confidential to protect the integrity of the deliberative process.

We wish to thank the following individuals for their participation in the review of this report: David Bodde, University of Missouri, Kansas City; Elisabeth Drake, Massachusetts Institute of Technology; Seth Dunn, Worldwatch

Institute; John Kaslow, EPRI Consultants, Inc.; Karl Rábago, Rocky Mountain Institute; Maxine Savitz, AlliedSignal, Inc.; Raymond Viskanta, Purdue University; and Carl Weinberg, Weinberg Associates. While these individuals provided constructive comments and suggestions, responsibility for the final content of this report rests solely with the authoring committee and the NRC.

Contents

EXECUTIVE SUMMARY 1

1 INTRODUCTION 10
Origin of the Study, 11
Office of Energy Efficiency and Renewable Energy, 11
Scope and Organization of the Report, 11

2 ROLE OF RENEWABLE SOURCES OF ENERGY 13
Energy Research and Development, 13
Government Support for Energy Research and Development, 15
Budget for the Office of Power Technologies, 17
Forces of Change, 18
Evolution of Research on Renewable Energy, 22
Strategic Rationale for the Office of Power Technologies' Programs, 24
Conclusions from Recent Energy Studies, 25
Comprehensive National Energy Strategy, 25
References, 26

3 ASSESSMENTS OF INDIVIDUAL PROGRAMS 28
Biopower Program, 29
Hydrogen Research Program, 38
Hydropower Program, 44
Geothermal Energy Program, 50
Concentrating Solar Power Program, 57
Solar Photovoltaics Program, 66

Wind Energy Program, 73
Crosscutting Programs, 78
Restructuring of the Electric Utility Industry, 79
Distributed Energy Resources, 83
International Issues, 88
References, 89

4   OVERALL ASSESSMENT OF THE OFFICE OF POWER
TECHNOLOGIES   92
General Findings, 92
Recommendations for the Overall Program, 99

APPENDIXES
  A Biographical Sketches of Committee Members   103
  B Committee Meetings and Activities   108
  C Summary of Recent Studies   113

ACRONYMS   123

Tables and Figures

TABLES

2-1 Funding for Renewable Power Technology Programs, 19

3-1 Long-Term Goals for Photovoltaic Technologies, 66
3-2 State Funding for Renewable Energy Development and Deployment, 81

FIGURES

2-1 Funding for renewable power technologies, FY95 to FY99, 18
2-2 Funding for crosscutting programs, FY95 to FY99, 20

3-1 Solar-trough system, 58
3-2 Dish/engine system, 59
3-3 Power-tower system, 60

Executive Summary

This report is the result of a study by the National Research Council (NRC) Committee for the Programmatic Review of the Office of Power Technologies (OPT) review of the U.S. Department of Energy's (DOE) Office of Power Technologies and its research and development (R&D) programs. The OPT, which is part of the Office of Energy Efficiency and Renewable Energy, conducts R&D programs for the production of electricity from renewable energy sources. Some of these programs are focused on photovoltaic, wind, solar thermal, geothermal, biopower, and hydroelectric energy technologies; others are focused on energy storage, electric transmission (including superconductivity), and hydrogen technologies. A recent modest initiative is focused on distributed power-generation technologies. In this study, the committee collected information and reviewed the activities of each of OPT's programs.

This Executive Summary presents the committee's recommendations for OPT as a whole and major recommendations for individual OPT programs. More detailed findings and recommendations for the individual programs can be found in Chapter 3.

OVERALL REVIEW OF THE OFFICE OF POWER TECHNOLOGIES

In order to meet both short-term and long-term energy needs, DOE must remain attentive to the constantly changing circumstances for new technologies. In the last decade or so, globalization of the economy has increased, priorities in Congress have changed, and the wholesale electric power-generation sector in many parts of the country has undergone restructuring. DOE's goal of providing options for the future energy needs of the United States requires a constant awareness of the effects of these changes on the nation's energy future and on R&D into renewable energy technologies.

The DOE was formed in the late 1970s in the wake of the oil embargo of 1973–1974, the resulting energy crisis, and sharp increases in energy prices. The focus of federal energy programs at that time was on reducing dependence on foreign supplies of oil and conserving the limited supplies of domestic oil and natural gas. These national security considerations led to a drive for energy reliability and security through increases in supply and control or reductions in demand. As the contributions of the energy sector to the U.S economy have become more apparent and the international market for U.S. energy technologies has grown, economic competitiveness has become a major goal. At the same time, environmental concerns, such as air quality and global climate change, have also emerged. R&D on renewable energy technology is now part of an overall approach to providing for clean, affordable energy, which is vital to the current and future well-being of the United States.

Substantial improvements in performance and reductions in cost of renewable energy technologies have been made. In fact, most of DOE's goals and objectives for cost and technical performance for renewable energy technologies have been met or exceeded, and the advantages and disadvantages of the various technologies are now well understood. However, renewable energy technologies have not met DOE's deployment goals. As a result, the use of renewable energy technologies in the U.S. economy is still limited.

Overall, the OPT's deployment goals for renewable technologies are based on unreasonable expectations and unrealistic promises. OPT has not developed the policies or resources needed to achieve its goals in an increasingly competitive electricity market, in which electricity can be generated relatively cheaply from conventional sources, such as natural gas and coal. Significant challenges will have to be overcome for renewable energy technologies to be competitive in a market in which the traditional customer (the utility industry) for the technologies under development is rapidly disappearing and is being replaced by diverse agents building and operating their own facilities.

Many experts believe that this distributed power generation will create opportunities for generating electricity in small units close to the users (e.g., at household, neighborhood, business, industry, or commercial locations). The trend toward smaller scale, more "distributed" generation technologies presents both challenges and opportunities for renewable energy technologies. Beyond studies of distribution systems, OPT will also have to address the relationship of each technology to the changing power grid. Other reasons for the lack of success in deployment of renewable energy technologies reflect changing national priorities and the changing role of DOE. Although deployment of renewable energy technologies domestically is included in DOE's overall goals, it has not been consistently funded by Congress. The international market will also offer substantial opportunities in the next few decades, especially in countries with high electricity prices and in regions that do not have transmission grids. A large potential market for many of the technologies under development by OPT is in the so-called "village power" markets, which are spread across the countryside of many developing countries where access to power grids is limited or nonexistent.

In this country, efforts to balance the national budget in the 1990s have constrained discretionary funds for energy R&D. Competing national needs, relatively stable, even declining, energy prices, and the absence of a sense of crisis have lessened the public focus on energy issues. As a result, DOE programs and staff have been cut back, and fewer new people are being brought in. As the DOE workforce ages and technical managers retire or leave the department, experienced, technical leadership declines.

Because only a small portion (20 percent) of DOE's total budget is currently directed toward energy R&D, strategic planning of energy R&D also receives proportionately less attention from top levels of DOE management. This lack of strategic thinking has led to OPT's lack of strategic focus. OPT's programs are not well integrated or coordinated but have operated as relatively separate groups with no common policy focus. This deficiency is especially apparent in light of the significant changes in the electric power industry. The committee recognizes the value of having separate technology groups striving to meet their own goals and in fostering competition of a sort. Although the objectives of individual programs are reasonably well thought out, they have not been considered in the overall context of OPT's goals or in light of the changing needs of the electric power sector. However, the committee believes that stronger OPT leadership and the formation of crosscutting teams could help OPT identify synergies among these programs. OPT also lacks a well-defined structure for linking its technology development programs to other R&D programs, such as programs in the DOE's Office of Science and other engineering research programs in and beyond DOE.

The committee is pleased to note that OPT has recently undertaken initiatives to develop a strategic focus and that DOE is in the process of analyzing the portfolio of DOE's energy R&D programs as a whole. The committee encourages OPT to survey all government and private-sector energy R&D and identify gaps that could be filled by renewable energy technologies, especially for collective or public-good benefits. OPT's strategy should also be integrated into and coordinated with DOE's overall energy R&D programs.

OPT must focus on reenergizing the strategic process and bringing its programs back on course in response to the significantly changed environment. In the past, DOE could easily "stay connected" to the thinking and planning of the electric industry, which was homogeneous and "open." The emergence of a competitive supply sector with different economic drivers, technology risk profiles, and commercial strategies has dramatically changed the situation and increased the need for strategic planning by OPT. New energy suppliers have competitive and overlapping interests. DOE's challenge now is to engage these new suppliers, as well as conventional suppliers, in future energy decisions.

Because of the ongoing restructuring of the U.S. electric power and energy industries in general and substantial reductions in R&D expenditures, state and federal governments now have even more reason to increase their involvement in energy R&D. However, at the time of the committee's review, OPT had paid little attention to a coherent roadmapping exercise that would include technical objectives and critical barriers to be overcome, a program for achieving objectives and setting priorities, the establishment of budget requirements, or the development of contingency plans to cope with uncertain budgets.

Although it appears that the individual programs have identified critical barriers to the development of individual technologies, OPT has not developed a systems analysis framework for examining the existing and emerging electric power system in detail and identifying the contribution (e.g., baseload, intermediate load, peaking, hybrid, etc.) that renewable energy technologies are most likely to make.

During the committee's study, OPT had begun planning a roadmapping exercise as part of its strategic planning initiative. In conjunction with this roadmapping exercise, and along with a deeper understanding of the factors that will contribute to the successful development and deployment of various technologies, OPT will have to develop criteria to help determine research priorities and the role of the public and private sectors in developing new renewable energy technologies. These priorities can then be used for the systematic direction of federal expenditures. Full life-cycle systems assessments and comparisons would be helpful for setting these priorities.

RECOMMENDATIONS FOR THE OVERALL PROGRAM

Recommendation. The committee encourages and recommends that the Office of Power Technologies (OPT) continue the roadmapping exercise and strategic plan it has initiated. Both the road map and the strategic plan should be consistent with the Comprehensive National Energy Strategy developed by the U.S. Department of Energy. The OPT strategic plan should be developed in collaboration with other agencies and sectors and should be integrated with a society-wide assessment of current activities by government agencies and private industry. The road map should distinguish between (1) R&D activities that promise to provide collective or public benefits and, therefore, require public oversight and (2) complementary R&D activities that primarily promise private benefits and can be left to the private sector. The roadmapping process should include an evaluation of how the technologies under development by OPT could contribute to the evolving electric power supply system, an identification of barriers to technical and market success, estimates of costs for reaching important milestones, and clarifications of federal priorities for development under budget constraints. Based on the road map, some new programs may be developed, some existing programs may be expanded, and existing programs that do not fit OPT's priorities and guidelines may be eliminated.

Recommendation. The Office of Power Technologies (OPT) should develop criteria, a rationale, and a systematic process for selecting research that should receive federal support in light of private sector and state-level activities. OPT should take advantage of the opportunity created by the restructuring of the electricity market to coordinate its activities with state-level renewable energy programs and assist them in implementing the results of OPT programs and promoting the deployment of OPT-developed technologies.

Recommendation. The Office of Power Technologies (OPT) should develop a robust rationale for its portfolio of renewable energy technology projects that will lead to a sustainable, cost-effective energy supply system for domestic and international markets. OPT in general, as well as individual OPT programs, should de-emphasize optimistic, short-term deployment goals as the metrics for defining success. The objectives should be the development of a sound science and engineering base, decreases in cost, improvements in technical performance, and the development of technologies that meet the needs of the marketplace. As technologies approach a level of readiness for the market, deployment strategies should be developed in cooperation with private sector agents, as appropriate, and higher policy levels in the U.S. Department of Energy.

Recommendation. The Office of Power Technologies should develop a systematic process for selecting specific research and development programs. The viewpoints of stakeholders should be considered in the development of selection criteria.

Recommendation. The U.S. Department of Energy should take advantage of existing government policies to promote the use of renewable energy technologies for electric power production by encouraging a public demand for "green power."

Recommendation. The Office of Power Technologies should focus more on integrating its programs, identifying common needs and opportunities for research, and clarifying how the individual programs can further their objectives. Benchmarking and other planning techniques used by industry could be adapted for measuring progress and selecting priorities. The challenges posed by the restructuring of the electric power industry, the use of distributed resource technologies, the need for storage technologies for many intermittent renewable technologies, and opportunities in the international market could be the integrating themes. One mechanism for facilitating integration among the individual programs would be to establish crosscutting teams to identify enabling opportunities and critical roadblocks and/or barriers to the development of technologies.

Recommendation. The Office of Power Technologies (OPT) should consider changing its organization and technology thrusts in several ways. Although the Hydrogen Research Program and work on superconductivity have important ramifications for the long term (and should be supported by the federal government), they should not be evaluated in the same way as emerging energy conversion technologies, such as photovoltaics or biopower. Hydrogen has energy carrier and/or storage capabilities that have long-term potential. OPT should develop a clear strategy for supporting long-term research.

Recommendation. The Office of Power Technologies should develop a clear strategy for the development of mechanical, electrical, or chemical storage technologies. Storage requirements for intermittent technologies should be considered in the context of the overall energy supply system. Today, natural gas turbines and pumped hydroelectric power can be used to provide supplemental energy. But promising "clean" energy carriers for the future (e.g., electricity and hydrogen) will require improved energy storage technologies. A breakthrough in either storage technology could strongly influence the future energy infrastructure.

Recommendation. The U.S. Department of Energy should establish a dedicated office to deal with distributed power systems. Whether or not this office is located in the Office of Power Technologies (OPT), its activities should be integrated with those of OPT.

Recommendation. The U.S. Department of Energy (DOE) should assess the effects of restructuring on the nation's electricity distribution system. DOE should provide support for research on distribution system behavior, operation, and control as a basis for assessing the effects of restructuring on electricity distribution systems. An understanding of these issues will be critical to the implementation of distributed generation technologies (which is the goal of OPT's programs). DOE should investigate the integration of distributed generation technologies into the evolving system. This investigation should be strategically coupled with the OPT program and with related activities in the building, transportation, and industrial sectors.

Recommendation. The U.S. Department of Energy (DOE) should provide funds for the direct support of graduate students through a DOE fellowship program leading to an advanced degree related to renewable energy research and development. This would ensure that an adequate supply of scientific and energy talent is available to the emerging industry and that new and inventive ideas continue to flow into the program.

Recommendation. The Office of Power Technologies (OPT) should institute a process for regular external peer reviews (at least every two years) of its proposed and ongoing projects and programs, as well as its overall goals. As part of the review process, OPT should publicly report how it responds to recommendations from external reviews.

Recommendation. Every Office of Power Technologies program should evaluate its resource assessment needs and should fund them accordingly. Resource assessments should be made in cooperation with the appropriate state agencies.

RECOMMENDATIONS FOR INDIVIDUAL PROGRAMS

In this section, key recommendations for individual OPT programs are summarized. Detailed reviews of the individual programs appear in Chapter 3, which also contains a wider variety of findings and recommendations for each program.

Biopower Program

DOE should look into establishing a center of excellence for bioenergy to bridge internal gaps in the Office of Energy Efficiency and Renewable Energy and create a strategic partnership with the U.S. Department of Agriculture for the development of crops and biobased products. This center should also be responsible for coordinating basic research activities in bioengineering by the Office of Science and the National Science Foundation and biosequestration activities by the Office of Fossil Energy. DOE should highlight the role of waste feedstocks in the current and future biopower market, educating stakeholders about environmental and market advantages. Education will be critical to the early success of biopower. Recently, bioenergy has become a major initiative of the Clinton administration, and OPT should position itself to play an active role in this initiative.

Hydrogen Research Program

The Hydrogen Research Program should be reoriented with a longer-term perspective and a stronger emphasis on the production of hydrogen from renewable resources (i.e., photolysis, biomass, and biological processes), the coupling of electrolysis with renewable energy generation, and distributed storage. OPT's Hydrogen Research Program should be coordinated with other elements in DOE, such as the Office of Transportation Technology, the Office of Fossil Energy, and the Office of Science that also are involved in hydrogen and hydrogen-related research.

Hydropower Program

To promote the preservation of existing hydropower capacity, as well for future development, hydropower conversion technologies that have higher efficiencies and that cause less damage to fish populations will be necessary. OPT should develop more environmentally sustainable, low-head, hydraulic-energy conversion systems for use in run-of-river and tidal basins. The initial focus should be on integrated technology and resource assessment to quantify the potential of low-head resources. The program should also explore new engineering concepts.

Geothermal Energy Technologies Program

DOE should reactivate its programs for the development of advanced concepts for the long term. The first priority of these programs should be high-grade enhanced geothermal systems; the second priority should be lower grade, hot dry rock, and geopressured systems. DOE should then support field demonstrations of enhanced geothermal systems technology. Although several new sites have been proposed for these demonstrations, such as Clear Lake, California, and Roosevelt Hot Springs, Utah, OPT should also consider sites in lower grade areas to demonstrate the benefits of reservoir concepts to different conditions. OPT should increase its R&D on reservoir diagnostics and modeling, especially on methods of detecting and enhancing *in situ* permeability.

Concentrating Solar Power Program

OPT should limit or halt its R&D on power-tower and power-trough technologies because further refinements to these concepts will not further their deployment. OPT should assess the market prospects for solar dish/engine technologies to determine whether continued R&D is warranted.

Solar Photovoltaics Program

The top priority of the Solar Photovoltaics Program should be the development of sound manufacturing technologies for thin-film modules. Much more attention must be paid to moving this technology from the laboratory through integrated pilot-scale experiments to commercial-scale design. This will require much more engineering expertise. Most laboratory-scale experiments could, with very slight modifications, provide critical information for eventual commercial-scale design. The program should make a concerted effort to integrate fundamental research and basic engineering research.

Wind Energy Program

The Wind Energy Program's research on advanced wind turbine technology should focus on turbulent flow studies, durable materials to extend turbine life, blade efficiency, and higher efficiency operation in lower quality wind regimes.

The development of advanced controls and improved gearboxes appear to be well within the capabilities of industry. OPT should investigate the potential of the global wind energy market because overseas markets could be essential for a struggling U.S. industry. Special requirements in these markets may include power system integration and a demonstrated ability to operate under many different environmental conditions.

Distributed Resources

DOE should develop a technology road map for distributed power-generation technologies to define the role of, and program goals for, distributed power systems in restructured electricity markets. DOE could then define the potential benefits of expanded markets for distributed power technologies and provide an analysis for policy decisions on distributed power markets. DOE should facilitate the development of commercial, institutional, and regulatory standards to open national markets to distributed power technologies. Local variations in standards, ranging from building codes and environmental permits to utility practices and tariffs, will require national coordination. DOE should increase its efforts to develop interconnection standards and national energy strategies to address institutional and operating barriers to the deployment of new technologies.

Crosscutting Issues

Assessing the effects of, and responding to, the changes in the electricity sector require global resource assessments and the identification of alternative markets. Wind, geothermal, and biomass power technologies are all faced by transmission barriers—especially rules for independent system operations. The success of intermittent renewable energy technologies (e.g., wind, photovoltaics) will depend on energy storage technologies. As renewable energy technologies mature toward market viability, these and similar issues should be included in OPT's technology development programs. A strong commitment to the integration of renewable energy technologies into the broader energy economy through crosscutting functions could improve the chances of real-world success for all renewable energy technologies.

1

Introduction

In the early twentieth century, most of the electricity in the United States was generated in large steam boilers fed by fossil fuels and designed to capitalize on perceived economies of scale (i.e., spreading fixed costs over a larger output). Other technologies, such as nuclear and hydroelectric power, have since become appreciable contributors to the generation of electricity. Electricity can also be produced by kinetic energy from the movement of air (wind-electric power), photovoltaic devices that convert sunlight directly to electricity, thermal energy to heat fluids that drive electric generators, and the conversion of solar energy into living material (biomass), which can be used as an energy source as well as a source of materials and food. If the sources of the energy are not consumed in the course of generating electricity, these processes are considered *renewable*.

Using the heat in the earth to generate electricity (geothermal power) is also considered a renewable energy process. Even though local geothermal energy will be depleted over time, given the abundance and magnitude of geothermal heat worldwide, geothermal power can be regarded as "renewable." Biomass is considered renewable because it is derived from solar energy. Because biomass takes carbon dioxide from the atmosphere as it grows, its combustion does not cause a net increase in atmospheric carbon dioxide (unless fossil fuels are used in the growing, processing, or transportation of biomass). When biomass wastes from forestry or agricultural activities are burned, the reduced need for waste disposal is an added benefit. In fact, experiments are under way on some quick growing crops that could be dedicated feedstocks for biopower projects.

The advantages of most renewable processes are low-cost or no-cost primary energy source (e.g., sunlight, wind, or geothermal energy), continuing availability, and little or no addition of greenhouse gases to the atmosphere. Despite these advantages, renewable energy processes must overcome substantial economic and other barriers to commercialization. Renewable energy technologies, which are in various stages of development, are the main focus of the U.S. Department of Energy's (DOE's) Office of Power Technologies (OPT), and the focus of this report.

ORIGIN OF THE STUDY

In response to a request from OPT for an independent review of its programs, the National Research Council formed the Committee for the Programmatic Review of the Office of Power Technologies (see Appendix A for biographical information). A Statement of Task was developed in consultation with OPT and its parent office, the DOE Office of Energy Efficiency and Renewable Energy (EERE), to conduct a programmatic review of OPT and recommend ways to strengthen the office and its programs.

OFFICE OF ENERGY EFFICIENCY AND RENEWABLE ENERGY

The EERE is responsible for developing cost-effective energy efficiency and renewable energy technologies that will protect the environment and support the nation's economic competitiveness. This goal is carried forward partly by OPT's programs to improve the cost and performance of renewable energy technologies. Working with industry through cost-shared technology development partnerships, OPT's research and development (R&D) is focused on solar-photovoltaic and solar-thermal power, biomass power, wind power, geothermal power, and hydroelectric power. OPT is also conducting R&D on advanced transmission and distribution technologies, energy storage, and hydrogen and is considering how renewable energy technologies can be used for the distributed generation of electric power. OPT's programs vary in size: the photovoltaics program has a budget of about $60 million per year; the hydroelectric power program and others have budgets of only a few million dollars per year. Both federal and private sector involvements will be crucial to the successful deployment of the developed technologies. DOE's goal is to facilitate deployment by using market mechanisms and by building partnerships with industry groups and state governments.

SCOPE AND ORGANIZATION OF THE REPORT

The committee's Statement of Task is reprinted below:

The National Research Council committee appointed to conduct this study will undertake a broad programmatic review of the OPT program. The review will be conducted in the context of the broader energy economy and in light of opportunities to leverage and coordinate activities among the eight programs within OPT as well as with energy R&D programs outside OPT. The review will address the eight programs in OPT: wind, photovoltaics, concentrated solar power, geothermal energy, hydropower, electrical systems and storage, biomass power, and hydrogen. The review will broadly consider programmatic issues such as:

- the goals of the programs and of OPT as a whole (especially in light of the current energy economy, restructuring in the electric power industry, and of recent energy R&D studies),
- processes for developing program plans, choosing R&D projects, monitoring progress, and directing program efforts & resources,
- the balance of short term vs. long-term R&D and the appropriateness of the technical directions being pursued,
- strategies for leveraging among the programs within OPT, other parts of DOE, other federal agencies, the private sector, and
- strategies for deployment.

The committee will prepare a report summarizing the major strengths and weaknesses of each of the OPT programs and make recommendations, if necessary, that in the judgment of the committee, would strengthen the office and its programs.

In response to requests from OPT and EERE, the committee has made recommendations for OPT as a whole, attempting to identify crosscutting themes as it reviewed OPT's individual programs. Experts were invited to make presentations and to join in discussions of OPT programs at committee meetings (see Appendix B).

The suggestions for improving OPT in this report are offered in the context of the current and projected challenges facing the United States. Because environmental issues and concerns about climate change are international, these issues are also relevant to the domestic energy picture. They are discussed in the context of their implications for OPT's renewable energy programs. The background and larger context of OPT's R&D programs are discussed in Chapter 2. Chapter 3 includes the committee's comments, reviews, and recommendations for each of OPT's programs. Chapter 4 includes the committee's findings and recommendations for OPT as a whole.

2

Role of Renewable Sources of Energy

As the new century begins, the United States and most other developed economies are faced with formidable challenges to ensuring that secure, affordable, and environmentally acceptable energy sources will be available to contribute to economic growth and improvements in the quality of life. Many domestic and global factors must be considered in determining and carrying out R&D on new technologies that can help meet these goals. This chapter provides a brief introduction to the key factors and issues the committee considered in its deliberations. Recommendations for helping OPT refine its strategic plans and define its role in delivering the next generation of advanced renewable energy technologies appear in subsequent chapters.

ENERGY RESEARCH AND DEVELOPMENT

The production and consumption of fuels and electricity have comprised a major sector of the U.S. economy since the industrial revolution. Energy is vital to virtually all components of the U.S. economy. In 1996, for example, expenditures for electricity in the United States reached $214 billion. This electricity was delivered by the power-generation industry, which is perennially the most capital-intensive sector in the economy (EIA, 1999). Even though the structure of the U.S. economy has changed dramatically over the last two decades from an economy based on heavy industry to one much more dependent on information and services, the role of energy, especially electricity, is still vital. Indeed, the availability of affordable, environmentally acceptable energy is central to the nation's economic well being and quality of life.

In fact, in the wake of growing information, service, and other light industrial sectors, as well as electrification of some industrial sectors (e.g., the steel industry), the economy is becoming more electricity intensive at a faster rate than in the past. In addition, a higher premium is now being placed on the quality of the electricity supply. Hence, in many economic sectors, low energy costs and a highly reliable supply of electricity have become crucial.

The production and consumption of energy in the United States and other modern economies also have significant national and international repercussions for the environment. Energy systems for producing electricity raise special concerns, such as the management of radioactive wastes from nuclear power plants and the management of emissions from the combustion of fossil or biomass fuels. Most scientists fear that the accumulation of greenhouse gases (e.g., carbon dioxide) in the earth's atmosphere, principally from the burning of fossil fuels, may lead to substantial climate changes. Slowing (or reducing) the buildup of greenhouse gases from the energy sector, both in this country and abroad, would require a substantial reduction in the carbon intensity of the world's energy system. In other words, the system would have to change to energy technologies that do not use fossil fuels to generate electricity, technologies that generate electricity from fossil fuels much more efficiently, or technologies that improve the efficient end-use of energy. Most likely, all three will be necessary. In addition, DOE is investigating options for continuing the interim use of fossil fuels with either carbon removal or capture and sequestration.

The involvement of the federal government in the energy sector in the last four decades has increased for reasons of national and energy security, economic vitality and international competitiveness, and environmental quality (including potential climate change). Federal involvement has included extensive programs in the development of energy technologies and involved substantial expenditures on R&D.

Before the oil embargo in 1974, most of the federal government's efforts to promote new energy supply technologies were carried out through the Atomic Energy Commission, with the development of nuclear energy, and the U.S. Department of the Interior, with the development of fossil fuels. Formal programs on energy efficiency or renewable energy technologies were rare before the mid-1970s. One of the earliest programs was a National Science Foundation (NSF) program, Research Applied to National Needs, which investigated alternative sources of energy. Following the oil price shocks of the 1970s, the short-lived Energy Research and Development Administration became part of a new cabinet-level department, the DOE, which has since become the lead agency for federal R&D on energy technologies, although other agencies also have relevant programs. DOE's energy R&D program is now approaching the quarter-century milestone. The NSF also has a large number of basic energy R&D programs.

Many aspects of the global energy economy are uncertain, as has been demonstrated by events such as disruptive energy price shocks, the emergence of environmental issues, and the explosive growth in the energy demands of developing countries. The U.S. federal R&D program was conceived initially as an investment in a portfolio of technological options to help the United States cope with uncertain future energy supplies, national security, and environmental circumstances. The underlying rationale for a federal R&D program in energy, especially renewable energy technologies, has evolved considerably in both scale and scope.

The committee focused on two aspects of this evolution: determining how well the OPT program has adapted to the changing economic, geopolitical, and environmental circumstances; and determining if OPT has established a process for monitoring circumstances to ensure that its programs are matched with anticipated needs.

GOVERNMENT SUPPORT FOR ENERGY RESEARCH AND DEVELOPMENT

Prior to the 1970s, the prevailing view was that the private sector was the appropriate place for the development of energy technologies. However, economists are quick to point out that there has been inadequate or little private support for R&D that is itself a "public good" (i.e., when results of the research will become widely known but the beneficiaries are uncertain) or when the primary initial beneficiary of the R&D is an activity with a large public component, such as national defense, improvements in basic infrastructure (e.g., roads, telecommunications, or the power and natural gas transmission and distribution grids), or activities that improve environmental quality but do not generally translate into the normal functioning of economic markets.

In the spectrum of R&D activities, those customarily described as basic research (e.g., clarifying the fundamentals of combustion, solar radiation, or nuclear fusion) are often viewed as public goods because the outcomes often cannot be anticipated, let alone the beneficiaries identified. Hence, basic research must often be supported by government. In some cases, if the outcome may affect national security, for example, the government may choose to undertake an investigation either to accelerate or limit the dissemination of results.

Firms in most competitive industries are unwilling to undertake basic research if the probability of an outcome beneficial to them is low or difficult to predict and thus cannot be translated directly into shareholder value. This is certainly true of most basic research on energy. In addition, if a firm foresees that it would have difficulty maintaining the advantage of the research, it will demur; this is becoming increasingly common as the economy becomes more competitive and globally connected. Hence, most basic research must be supported directly by government or supported indirectly through universities or other research institutions.

Applied R&D is more often the focus of private sector funds. However, because not all of the benefits of improved energy technologies (e.g., environmental benefits) can be captured by private interests, R&D in this area has been generally underfunded. This is the traditional justification for government sponsorship of R&D for less environmentally intrusive forms of energy conversion (e.g., clean combustion, solar energy technologies, fuel cells), although technology development has been pursued by private interests to meet new regulatory standards or, more recently, to avoid effluent fees on the emissions of pollutants.

R&D on innovations that enhance a shared, facilitating mechanism (even though the users may derive some private benefit), such as improved traffic control on road networks or improved power grid operation, also require government support. In these cases, R&D is sometimes financed by charges imposed on the ultimate private beneficiaries who do not generally voluntarily agree to fund the R&D. Other R&D that requires government support is focused on the efficient consumption of common-pool, nonrenewable resources (e.g., oil and gas fields not owned by a single owner) or renewable resources approaching the threshold of extinction (e.g., biomass feedstocks).

Another rationale for public intervention in applied R&D is in an industry with an institutional structure that may underallocate or misallocate the benefits (e.g., if an industry exhibits monopolistic behavior or if a national security interest might be affected by the involvement of a multinational enterprise). In these instances, the public is unlikely to benefit fairly from innovation without public intervention and/or support. For example, all other things being equal, an electric utility with a local monopoly franchise and excess generating capacity is not likely to have a compelling reason to invest heavily in R&D on new generation technologies, even though for many reasons that utility's customers might benefit.

Finally, government also becomes involved in industrial R&D when a national priority or concern has been perceived. Because profitability is a prime consideration of private sector operations, industry has few incentives to think of anything but short-term returns to satisfy stockholders. With competitors waiting to capture market share, few businesses will risk resources to develop products that do not promise immediate returns. In the energy sector in particular, the trend of the last decade toward more competitive markets has led to a marked decline in industry-sponsored research, and even less is expected in the future. Company and other sources of industrial R&D declined from $2.4 billion in 1987 to $884 million in 1997 (NSF, 1997). This trend has persisted despite the general recognition that U.S. industry, in many sectors including the energy sector, frequently enjoys a competitive edge in global markets because of the results of R&D. For reasons that have less to do with strategic positioning of the federal R&D program and more to do with competing priorities and budget cutbacks, the decline in industrial spending on research has been mirrored by a similar drop in federal government-sponsored research. Overall, federal funds for energy R&D fell throughout the 1980s and 1990s, from $1.2 billion in 1987 to $756 million in 1997 (NSF, 1997). In fact, spending for energy R&D has declined significantly in the last 20 years across the industrialized world. The notable exception to this trend is Japan. While energy R&D spending declined by 58 percent in the United States between 1980 and 1995 and by some 85 percent in Germany, Japan increased its energy R&D investment by some 20 percent in the same period. Some have argued that these cutbacks are detrimental to U.S. energy security and will reduce the capacity of the energy sector to innovate and respond to emerging risks on the international fronts, such as global climate change (Margolis and Kammen, 1999).

With fewer dollars available overall, government has increased its attempts to work cooperatively with private industry and others to define attainable long-term R&D goals. However, the decline in federal sponsorship of R&D is only one of the forces shaping the landscape for the development of the next-generation renewable energy technologies. Some other influences are the structure of the economy, the maturity of technologies developed to date, the prevailing attitudes toward regulation, the growing complexity of environmental issues, and change from a bipolar world with two superpowers to a world of regional hot spots. In the following sections, many of these changes are described, and their implications for R&D on renewable energy are outlined.

BUDGET FOR THE OFFICE OF POWER TECHNOLOGIES

Funds for OPT are included in the congressional appropriations for energy and water. Much of the spending is directed as obligated funds for areas such as high energy or nuclear physics. Energy supply activities, including program obligations for solar and renewable energy and nuclear energy R&D, are included in the obligation for energy research analyses. The budget line item for solar and renewable energy encompasses the majority of OPT's programs.

Thus, appropriations for OPT's programs are managed in great detail by Congress, and, as a result, the management of OPT is often as much involved with political and budgetary processes as with R&D technology issues. The total budget for OPT programs was just over $300 million in fiscal year (FY) 1995 but was cut back markedly the next year. In recent years, the budget for OPT has rebounded somewhat (Figure 2-1), which shows the historical mix of funds for renewable power technologies.

For budgetary purposes, OPT's programs have been divided into three functional areas: renewable power technologies; power delivery technologies; and cross-program activities. Funding for FY00 and FY01 are shown in Table 2-1. The line item for cross-program activities includes the following: solar program support; international issues; climate challenge; renewable energy resources for Native Americans; and federal buildings/remote power.

The funding for crosscutting programs referred to in this report (included in the power delivery category in Table 2-1) is shown for FY95 through FY99 in Figure 2-2.

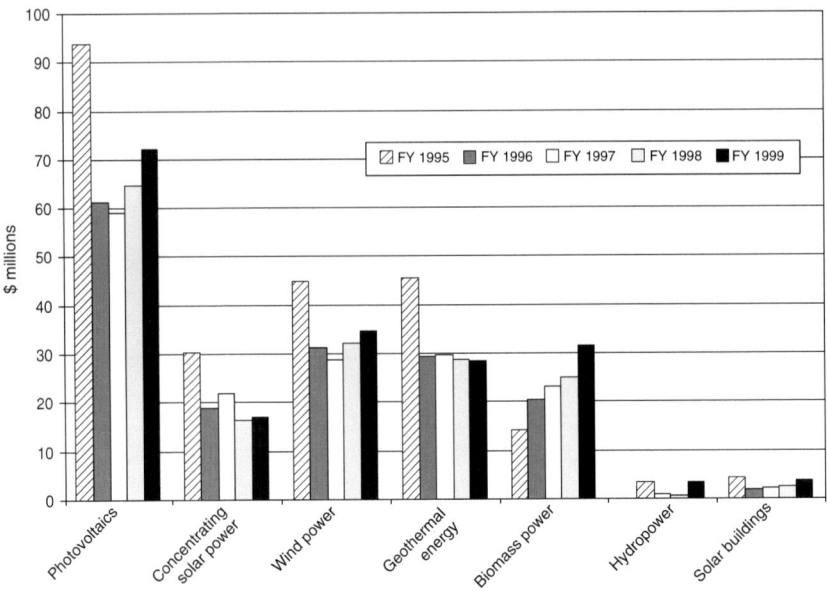

FIGURE 2-1 Funding for renewable power technologies, FY95 to FY99. Source: DOE, 1999.

FORCES OF CHANGE

Although the development of renewable energy sources has been an integral part of U.S. energy policy since the early 1970s, the motivation for using renewable energy sources has changed considerably. In the 1970s, in the wake of the oil embargo, policy was driven by energy security concerns. Today, one could argue that environmental concerns are dominant. Therefore, although energy security is still a long-term goal for the development of alternative energy technologies, the urgency of the 1970s no longer prevails. The current oil market is much more diversified, the availability of natural gas has dramatically expanded, and the efficiency of energy use in the economy has considerably improved. Other changes in the economic, environmental, and geopolitical situations of energy markets have also had profound effects on energy supply and demand. A variety of environmental, economic, and security concerns have arisen for preserving and nurturing alternative energy options:

- **End of the Cold War.** The change from a world dominated by a bipolar struggle between the United States and the Soviet Union to one with more

TABLE 2-1 Funding for Renewable Power Technology Programs, FY00 and FY01

	FY00 Enacted ($ millions)	FY01 Request ($ millions)
Renewable Power Technologies		
Photovoltaics	65.9	82.0
Solar buildings	2.0	4.5
Concentrated solar power	15.2	15.0
Biopower	31.8	48.0
Geothermal power	23.6	27.0
Wind power	32.5	50.0
Hydropower	4.9	5.0
Power Delivery		
Superconductivity	31.4	32.0
Energy storage	3.4	5.0
Hydrogen	24.6	23.0
Transmission reliability/distributed power	3.0	11.0
Cross-Program Activities	16.7	31.6
Total (rounded)	255	334.6

Source: Dixon, 2000.

regional risks has precipitated sweeping geopolitical changes. As the sense of urgency about national security interests has diminished, the case for heavy expenditures in R&D in many areas, including energy, that had traditionally been based on national security objectives must now be rationalized in other ways. Formerly funded R&D must now compete with other policy imperatives, such as health care, for funding.

- **Globalization of trade, finance, and industry.** Revolutions in telecommunications and transportation in the last two decades have led to the development of global financial markets, dramatically increased trade in commodities and technologies, and led to the rapid growth of multinational enterprises and activities, including R&D. Mergers and acquisitions, for which overall cost savings are often cited as benefits, have also reduced funds for "discretionary" activities (such as R&D). Major mergers and acquisitions involving multinational firms have also changed the cast of players involved in the renewable energy technology business dramatically over the last decade. Moreover, as long as energy prices in the United States remain low, the early markets for many renewable technologies are likely to be overseas, especially in developing countries.

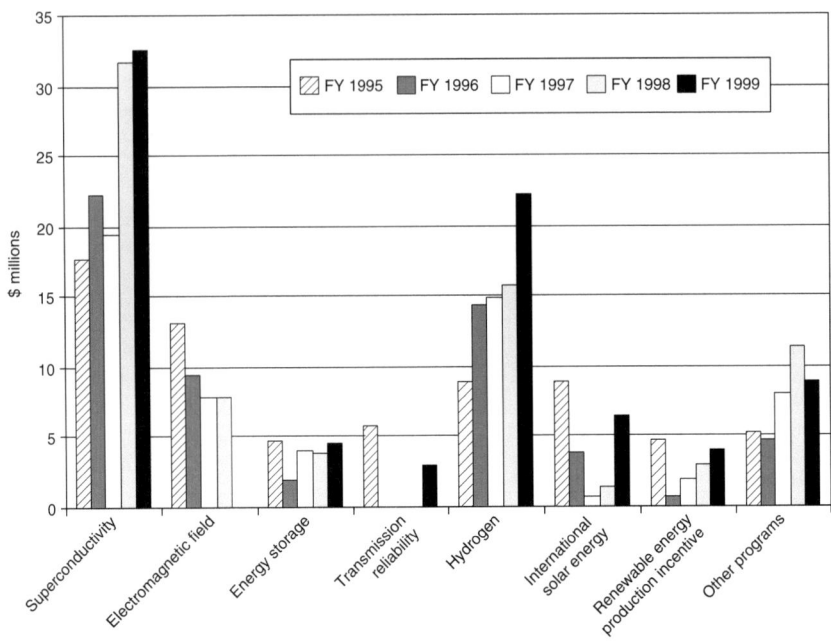

FIGURE 2-2 Funding for crosscutting programs, FY95 to FY99. Source: DOE, 1999.

- **Widespread adoption of market-based approaches to regulation.** In the last decade, successful experiments in economic deregulation and market-based environmental regulation have led to the adoption of similar strategies at all levels of government. Sweeping changes in federal and state legislation have changed the economic and environmental regulatory context of the energy business, especially the utility business.
- **Restructuring of the electric utility industry.** A prime example of the trend toward market-based regulation is the transformation of the electric utility industry to an industry centered on competitive markets for power generation and final markets in other areas. A competitive electricity market may, with appropriate policies, improve the deployment or renewable energy technologies in the long term; these changes have seriously undermined the short-term climate for the adoption of new technology in the electric power business by reducing available funds for industry investment in long-term power-generation alternatives.
- **Restructuring of the oil and gas sectors.** Sweeping changes in the oil and gas business over the last decade, including the evolution of futures

markets, the diversification of world oil supplies, and dramatic increases in the availability and discovery of new natural gas reserves, have led to persistently low oil and gas prices, which have greatly weakened incentives for the development of alternative energy supplies.

- **Increased role of state governments.** The diminishing federal role in energy R&D, coupled with the restructuring of electricity markets in many states, has actually increased the level of activity supported by some states in the development of renewable energy technologies. These state-sponsored programs are likely to have an important impact on the early commercial adoption of some renewable energy technologies in the United States. Renewable portfolio standards and/or systems benefits charges (SBCs) are included in the power industry restructuring in 13 states. In general, the restructuring of the power industry in the direction of more competitive markets has been driven by the potential economic benefits of lower costs and lower prices for electricity. SBC funds have been created by states to enable them to continue funding the public benefit programs that were originally implemented through utility rate structures. SBC funds generally include a component (usually a small component) for improving efficiency overall and developing renewable energy systems during the transition to a fully competitive market. Although SBCs are designed to last for a limited period of time, they do create a substantial challenge to the renewable energy technology community because the infusion by states of almost $1.6 billion through 2010 into technology development and deployment is an opportunity that is not likely to be repeated (Wiser et al., 1999; see Chapter 3). If state programs fail to achieve defined goals, it will be difficult to justify continuing the investment once the current financial incentives expire.
- **Improved understanding of the global environment.** Research on global environmental issues has begun to sharpen divisions in the policy debate about global climate change and other issues. Although many uncertainties remain in the science of global climate change, many uncertainties have been resolved or reduced over the past decade. For example, it is now generally accepted that the activities of human populations do affect the global environment, and the debate is shifting to how far-reaching those effects are (e.g., the consequences of a rise in the average temperature of the earth, which has changed weather and rainfall patterns; the incidence of extreme weather events; and rising sea levels). Other serious environmental concerns, such as emissions of sulfur, nitrogen oxides, fine particulates, mercury, and other toxic materials that affect air quality, are crucial for developing countries. Most of these are principally by-products of fossil fuel combustion.
- **Transition from deficit spending to surpluses in U.S. federal budgets.** The budget austerity that dominated discussions of federal R&D in the

1980s and much of the 1990s have given way to federal budget surpluses. Current constraints on R&D expenditures are influenced more by competition with other federal programs and priorities, such as health care, than by a desire to balance the budget.
- **Decrease in total energy R&D.** More than 80 percent of DOE's budget goes to areas other than energy R&D.
- **Erosion of the boundary between basic and applied research.** Innovations in industry are being made at all levels, and even across levels. Better information about industrial processes, more flexible technologies and materials, better process controls, and many other factors have blurred, and sometimes eliminated, the traditional distinction between basic and applied research. For example, fundamental discoveries in materials can make their way into applications, and the experiences of those applications can prompt new directions in basic research so quickly, that the traditional sequence breaks down. Rapid changes have fundamentally changed the selection of research directions and the way proposed projects fit into an integrated program. Thus, the old model of the linear progression of R&D is breaking down. Many renewable technologies will require introduction into the market to complete the development cycle with modular units suitable for mass production.

In the past decade, DOE's R&D program, including OPT in renewable energy, has not been able to adapt its strategic rationale for program activities and priorities in response to these changes. Even decisions about the R&D portfolio based on traditional factors, such as persistently low energy prices and uncertain and declining R&D budgets, are not reflected in R&D planning and priorities. As a result, OPT's overall program appears to be outdated, burdened by inertia, and suffering from a lack of clear direction.

Until recently, DOE was able to "stay connected" to the thinking and planning of the electric industry, which was rather homogeneous and "open." The emergence of a competitive supply sector with different economic drivers, technology risk profiles, and commercial strategies, has challenged DOE to engage new suppliers, as well as conventional suppliers, in future technology decisions.

EVOLUTION OF RESEARCH ON RENEWABLE ENERGY

In the 1970s and early 1980s, energy prices were expected to rise, and the opportunity for using natural gas in industry, especially for the production of electric power, was expected to be limited. Indeed, for nearly a decade, the Powerplant and Industrial Fuel Use Act prohibited natural gas from being used in industrial boilers or electric power generation. In the wake of the discovery of large resources of natural gas, that legislation was repealed in 1986.

Since then, the widespread availability of natural gas, considerably more efficient uses of energy, and numerous other technological improvements have all contributed to very low energy prices and relegated most alternatives to natural gas technologies to niche markets for the foreseeable future. Even though most renewable energy technologies have met or exceeded expectations with regard to performance and cost, renewables have not met deployment goals mostly because of the declining price of conventional electric power generation (Burtraw et al., 1999).

The goals, policies, and technology development programs established in the 1970s were intended to reduce U.S. dependency on foreign oil while conserving U.S. resources of oil and natural gas. A key piece of legislation enacted during this period was the Public Utility Regulatory Policies Act of 1978 (PURPA). Although it did not emerge from court challenges until 1983, PURPA encouraged the use of alternative energy technologies in electric power generation. The cogeneration of thermal energy and electrical power to improve the efficiency of fossil energy use in electric power generation, which was included almost as an afterthought, dominated the implementation of PURPA in the 1980s, even in California, the state with the most extensive deployment of renewable energy technologies. PURPA was also instrumental in accelerating the commercial deployment of renewable energy technologies in the 1980s, especially wind power technologies, but also geothermal technologies and, to a much lesser extent, solar-thermal technologies. Federal sponsorship of demonstration projects, federal and state tax credits and other subsidies (e.g., loan guarantees, low interest loans, and grants), the creation of so-called "standard offer" contracts for new projects to minimize the hassles of negotiating new projects, added costs to meet increasingly stringent regulatory requirements on traditional sources of power-generation (e.g., coal and nuclear energy), a financially struggling electric utility industry, a general expectation of continued increases in energy prices, and occasional geopolitical events that focused attention on energy security all contributed to the deployments of renewable energy technologies in the early 1980s.

At the same time, natural gas markets became increasingly deregulated, precipitating a number of important energy technology trends, such as dramatic improvements in exploration and drilling technologies and in the cost and performance of combustion turbine technologies for electric power generation. These trends fundamentally changed the U.S. energy outlook, resulting in persistently low natural gas prices and technology costs relative to other fuels and the current dominance of natural gas-combined cycle units for new electric power generation projects. The impact of low cost, modular natural gas units, and the expiration of many subsidy programs have resulted in a vastly diminished domestic renewable power industry.

Despite the dominance of natural gas in new energy markets, concerns about the long-term availability of clean domestic sources of energy are still the basis for the development of renewable energy technologies that would consume virtually no finite resources. However, the R&D agenda for renewable energy technologies has formidable cost and performance hurdles to overcome.

STRATEGIC RATIONALE FOR THE OFFICE OF POWER TECHNOLOGIES' PROGRAMS

The significant changes in energy markets and the restructuring of the industry in the last decade are not reflected in the strategic direction of OPT's technology programs for the next decade, which are still focused on the cost, performance, and security goals established during the late 1970s and 1980s. The following national energy interests should drive OPT's programs (Office of the President, 1997):

- competitive market entry
- environmentally sustainable energy supplies
- national energy security and the reliability of critical infrastructure

Competitive markets are creating new opportunities for particular technologies as commercial and residential buyers are beginning to purchase energy systems tailored to meet specific energy needs. Environmentally preferred and locally distributed energy supplies are examples of new products that could accelerate commercial demands for OPT-developed technologies. At the same time, uncertainties about interconnection pose significant barriers to the market entry of some new technologies. Establishing a strong domestic market for renewable energy technologies will also drive competitiveness in global markets. Industry investments are also being influenced by the addition of risk-mitigation strategies for carbon dioxide emissions to regional and local public health and safety regulations.

Increased energy imports make the diversity and security of energy supplies important national considerations. The restructuring of the electricity market has created both opportunities and challenges (e.g., redesign of the transmission and distribution grid to support reliability and competitive access goals) to the deployment of renewable energy technologies. If these technologies become a significant part of the domestic power generation mix, they would not only add to the diversity of energy supplies but would also add to the security of energy supplies.

In summary, the strategic drivers described in the section above should define the role of the OPT. Renewable energy technologies developed by OPT could increase the options for meeting national energy objectives, but OPT programs are not currently designed to meet these objectives. OPT has only recently begun to explore ways to redirect its programs to meet the strategic needs of the United States.

CONCLUSIONS FROM RECENT ENERGY STUDIES

A number of studies related to energy technologies and associated R&D have been conducted during the past few years (see brief summaries in Appendix C). A study in 1995 by the Task Force on Strategic Energy R&D recommended that DOE benchmark its R&D management practices against "best practices" elsewhere and develop an integrated strategic plan and process for energy R&D, including the establishment of priorities (DOE, 1995). *Technology Opportunities to Reduce U.S. Greenhouse Gas Emissions*, known as the Five-Laboratory Study, found that renewable energy technologies could have a major effect on the reduction of carbon emissions to the atmosphere, but mostly after 2010 (DOE, 1997). *Scenarios of U.S. Carbon Reductions*, the Eleven-Laboratory Study, also concluded that renewable energy technologies have a significant potential to reduce greenhouse gases to the atmosphere by displacing electricity generated by fossil fuels and that a national investment in R&D and demonstration over the next three decades would provide a portfolio of technologies that could significantly reduce greenhouse gas emissions (EERE, 1997).

A recent report by the President's Committee of Advisors on Science and Technology (PCAST) on federal R&D noted that, although costs have been reduced significantly, the primary challenge facing renewable energy technologies is the relatively high unit costs compared to unit costs using abundant fossil fuels. PCAST recommended significant increases in R&D budgets for renewable energy technologies to increase the probability of developing viable energy options to meet a variety of environmental challenges (PCAST, 1997). In a more recent report, PCAST identified significant international opportunities for renewable energy technologies and suggested that the development and deployment of renewable energy technologies be accelerated, especially technologies that might be appropriate for use in rural areas of developing countries (PCAST, 1999).

COMPREHENSIVE NATIONAL ENERGY STRATEGY

Despite existing regulations and possible international restraints, the U.S. energy strategy should reflect a balance between environmental concerns and industrial needs to encourage economic growth and decrease the environmental impacts of energy use. In cooperation with other federal agencies, and through a public hearing and comment process, DOE recently codified a Comprehensive National Energy Strategy (CNES) (DOE, 1998). The purpose of this policy plan is to address the major energy challenges facing the United States and to provide a basis for directing and guiding future action. The plan is based on the following five goals:

- Improve the efficiency of the energy system by making more productive use of energy resources to enhance overall economic performance while protecting the environment and advancing national security.
- Ensure against energy disruptions by protecting the U.S. economy from external threats of interrupted supplies or infrastructure failure.
- Promote energy production and use that reflect human health and environmental values thus improving health and local, regional, and global environmental quality.
- Expand future energy choices by continuing to pursue science and technology to provide future generations with a robust portfolio of clean, affordable sources of energy.
- Cooperate internationally to address global economic, security, and environmental concerns.

DOE's policy objectives are to focus attention on the importance of energy in the U.S. economy and national security, as well as to increase awareness of the environmental effects of using fossil fuels to produce energy. Thus, DOE hopes that public knowledge of how energy is supplied and used will encourage the efficient utilization of energy resources. In the following chapters the committee examines how well the OPT program portfolio furthers these policy objectives and how well the program plans can be adapted to changes in the global and domestic energy markets.

REFERENCES

Burtraw, D., J. Darmstadter, K. Palmer, and J. McVeigh. 1999. Renewable energy: winner, loser, or innocent victim? Resources 135: 9–13.

Dixon, R. 2000. Energy Efficiency and Renewable Energy in the Office of Power Technologies. Presentation by Robert K. Dixon, acting deputy assistant secretary for the Office of Power Technologies, to the Committee for the Programmatic Review of the Office of Power Technologies, U.S. Department of Energy, Washington, D.C., February 7, 2000.

DOE (U.S. Department of Energy). 1995. Report of the Task Force on Strategic Energy Research and Development. Washington, D.C.: Secretary of Energy Advisory Board, U.S. Department of Energy.

DOE. 1997. Technology Opportunities to Reduce U.S. Greenhouse Gas Emissions. Washington, D.C.: U.S. Department of Energy. Also available on line at: *http://www.ornl.gov/climate_change*

DOE. 1998. Comprehensive National Energy Strategy: National Energy Policy Plan. DOE/S-0124. Washington, D.C.: U.S. Department of Energy.

DOE. 1999. Available on line at: http://www.eren.doe.gov/power/budget.html

EIA (Energy Information Administration). 1999. Annual Energy Outlook 1999. Washington, D.C.: Energy Information Administration, U.S. Department of Energy.

EERE (Office of Energy Efficiency and Renewable Energy). 1997. Scenarios of U.S. Carbon Reductions: Potential Impacts of Energy Technologies by 2010 and Beyond. Washington, D.C.: U.S. Department of Energy.

Margolis, R., and D. Kammen. 1999. Underinvestment: the energy technology and R&D policy challenge. Science 285(5427): 690–692.
NSF (National Science Foundation). 1997. Survey of Industrial Research and Development. Arlington, Va.: National Science Foundation/Statistical Research Service.
Office of the President. 1997. Critical Foundations: Protecting America's Infrastructures. Report by the President's Council on Critical Infrastructure Protection. Washington, D.C.: Office of the President.
PCAST (President's Committee of Advisors on Science and Technology). 1997. Federal Energy Research and Development for the Twenty-First Century. Washington, D.C.: Executive Office of the President.
PCAST. 1999. Powerful Partnerships: The Federal Role in International Cooperation on Energy Innovation. Washington, D.C.: Executive Office of the President.
Wiser, R., K. Porter, and S. Clemmer. 1999. Emerging Markets for Wind Power: The Role of State Policies under Restructuring. In Proceedings WINDPOWER 1999. Available on CD-ROM only. American Wind Energy Association, 122 C Street, NW, 4th Floor, Washington, DC 20001.

3

Assessments of Individual Programs

The OPT is home to a diverse array of renewable energy technology programs geared to R&D of commercially viable systems that meet DOE's general goal of producing clean, affordable energy. OPT's R&D mission is described in the following statement (DOE, 1999a):

> A key strategy in accomplishing OPT's mission is to establish and maintain a renewable energy technology base. The OPT works with industry, state and local governments, universities, and the DOE's national laboratories to support aggressive research and development in photovoltaic, concentrating solar, wind, geothermal, hydropower, and biomass power technologies and systems. Much of this research is cost-shared with industry, whose contribution is typically 30 percent–50 percent of a total project budget, particularly for system hardware development and demonstration. Industry's willingness to share the cost of R&D indicates its belief in the market potential of these technologies and its commitment to commercialize them.

Thus, the goal of OPT is to develop promising renewable energy technologies to the point at which the private sector can evaluate their viability under anticipated market conditions. If warranted, industry will then assume a major responsibility for their deployment and commercialization.

Because OPT's programs are at different stages in the R&D and deployment cycle, conducting quantitative, comparative evaluations is difficult. In this chapter, the committee evaluates OPT's programs based on presentations by DOE program personnel, laboratory staff, and representatives of industry and academia, as well as the experience and personal knowledge of committee members. The committee considered many factors: the current state of market acceptance and private sector interest; cost and performance profiles; technology development track record (including gaps and perceived needs); prospects for continued improvement of the technology; and the likelihood of access to necessary resources.

The committee considered how changing circumstances have affected and how continuing change is likely to affect the deployment of new technologies; considerations include regulatory trends, developing international markets, policy influences, and improved understanding of the environmental consequences of technology.

Until now, electricity generation, transmission, and distribution in the United States favored large, central station systems to provide baseload capacity. But current conditions are creating a market favorable to smaller, modular, cleaner distributed-generation technologies. Economic criteria are becoming more stringent, and value considerations beyond the cost of power generation are beginning to influence the market.

Given the broad uncertainties in the forces that will shape our shared energy future, and the growing need worldwide for reliable, affordable, clean energy adaptable to varying local and regional circumstances, we must build flexibility and adaptability into our electrical energy systems. A variety of technology options will reduce the risk of "energy" surprises. Therefore, OPT's plans should be sensitive to the likelihood of change and uncertainties.

BIOPOWER PROGRAM

Plans and Goals

The Biopower Program is focused on advanced technologies for producing electricity from renewable biomass. DOE has also undertaken a bioenergy initiative to develop national partnerships with other federal agencies and the private sector. Integrated R&D on bioenergy will encompass existing R&D by DOE on transportation fuels, biomass power, forest products, and agricultural industry programs to encourage the development of a variety of fuels, power sources, chemicals, and other products (NRC, 1999; Reicher, 1998). Bioenergy has become a major initiative of the Clinton administration, and OPT should position itself to play an active role.

The goals of the Biopower Program closely match the goals defined in the Comprehensive National Energy Strategy (CNES) (DOE, 1998a) and overlaps the missions of the DOE's Office of Industrial Technologies (OIT) and the Office of Transportation Technologies (OTT). A number of activities in OIT are focused on the agricultural and forestry sectors, and OTT has an office focused on the production of liquid transportation fuels from biomass resources. The ultimate mission of all three is to create either new competitive businesses or increase the global competitiveness of existing industries.

Program Priorities

The Biopower Program has an aggressive agenda for developing and demonstrating advanced technologies to convert biomass resources to power. Currently, biomass power-generation projects are combustion-based, with gasification considered the technology of the future. By building partnerships through the Rural Development Initiative and sharing the risk of bringing new technologies to market in the major gasification projects, the research program had hoped to have an impact on the power market within the next 10 years. However, the wholesale changes in the electric utility industry have effectively created an opportunity for DOE to promote the use of biomass for electric power generation. State legislative and regulatory restructuring of the utility industry, driven by differential electricity prices and competition among states for new industries, has resulted in a new competitive market (EIA, 1999a).

As of April 1999, 23 states had enacted legislation or promulgated regulations establishing retail competition programs (DOE, 1999b). The customers for the technology developed in the Biopower Program are changing, along with the criteria for success.

According to the *Annual Energy Outlook 1999*, state renewable energy programs are expected to result in more than 630 megawatts of new capacity between now and 2011. Biomass is forecast to provide more than 130 megawatts of this new capacity (EIA, 1999b). In the transition period, substantial financial resources have become available at the state level (systems benefits charge funds) for promoting and implementing public benefit programs. Therefore, OPT has an opportunity to work with states to ensure that they are invested wisely. Public benefit programs are also focused on increasing energy efficiency, and in some states, the funds are being targeted to economic development by creating new business enterprises to meet the demand for renewable energy technology. This is an opportunity for OPT to coordinate its activities with state-level programs, helping the states to implement the results of their efforts and promote the deployment of OPT-developed technologies.

The Regional Biomass Program (a component of the Biopower Program) addresses the needs and concerns of the general public. With financial support from the Regional Biomass Program, a majority of states now have a staff person devoting at least some time to biomass issues. By carefully nurturing these liaisons, the Biopower Program could learn more about state-level issues related to the development of biopower and, in the long run, create allies at the state level. The federal government can also promote advanced biopower technologies by using them to provide electricity for federal government facilities.

Although state organizations will be the primary near-term partners for the deployment of renewable energy technologies, the development and commercialization of new technologies will require partnerships with research institutions and private industry. For these relationships to be productive, OPT will have to overcome many barriers. Some potential partners have cited the administrative burden of doing business with DOE, which, they say, takes time and resources away from the primary R&D effort as a barrier to their participation. DOE's annual budget uncertainties also make it difficult for contractors to plan ahead and can ultimately cost the project money.

Reaching the deployment targets for biopower technologies will require a concerted effort by OPT to change the perception that the combustion of wood, waste wood in particular, is not a "green" (i.e., environmentally friendly) technology. OPT will have to educate the environmental community and the public at large to the idea that the growth of energy crops and the use of waste feedstocks can have a variety of positive environmental effects. For example, an important attribute of biomass as a renewable feedstock is that it will have minimal, if any, impact on global climate change. Building support in the environmental community for the use of biopower will require focusing attention on strategies to maximize the role of biopower in mitigating global climate change without adversely affecting biodiversity because areas used for energy crops must be cleared periodically for harvesting (Beyea, 1999). Bioenergy systems can also potentially help protect watersheds. In fact, in some locations, the environmental benefits of biopower may be the driving force behind the initial establishment of bioenergy plantations. Convincing states and the public, however, will require sound research and analysis of the environmental consequences of various biopower strategies and the dissemination of the results to stakeholder groups (Peelle, 1999).

Research Issues

Partnerships with industry, national laboratories, and universities can provide the necessary skills to move technologies or concepts from fundamental research through the stages of development to the commercial market. Through well chosen partnerships, for example, research capabilities could be focused on meeting the needs of industry (PCAST, 1997). DOE's Bioenergy Initiative for coordinating DOE activities in this area could be extended to other federal agencies with similar interests, such as the U.S. Department of Agriculture (USDA). In addition, OPT could undertake an assessment of biomass resource end-uses from electric power generation to fuels and chemicals production and use the results to set priorities for the development of biomass resources for targeted end-uses. The Biopower Program could also promote the collateral benefits of biomass crops, such as providing wildlife habitat, reducing greenhouse gases, and others.

The major thrusts of the Biopower Program are focused on meeting three strategic targets: (1) increasing opportunities for rapid near-term deployment of cofiring biomass in existing boilers; (2) linking energy crop production and conversion via gasification and other advanced process technologies; and (3) establishing a role for biopower in the distributed power-generation market through the development of modular systems. Five primary technical barriers have been identified: biomass resource productivity; materials handling; biomass conversion; combustion contaminant reduction; and integration with current power-generating systems.

Research on modular systems is targeted to the needs of the distributed power-generation industry in the United States and international markets. Engineering design and prototype construction are planned for projects that successfully complete the feasibility stage. Because a large financial commitment will be necessary to bring multiple new technologies to the market, DOE could establish a center of excellence for bioenergy to house these projects. Based on budget presentations, the committee was unable to determine if the Biopower Program has considered the need for a long-term partnership with industry to develop the technologies selected for scale-up or the impact of such a partnership on the overall program budget.

Given the dramatic changes in the power generation business and the new customers for OPT technologies, the Biopower Program should focus on developing the biomass resource base, understanding infrastructure needs, and identifying market opportunities.

Commercial Prospects and Market Barriers

A core responsibility of government is to strengthen America's educational system in science and technology to achieve societal goals for the twenty-first century (OSTP, 1997). The creation of a strong market for business development in biopower technology will require a substantial investment in education and training. The greater the investment in postgraduate education, the higher the rate of formation of new firms (Reynolds et al., 1999). Public awareness of the technologies and applications of biopower systems will ultimately contribute to a sustainable biopower industry.

In the near future, however, the transition to a competitive market will keep the power industry in flux. The cofiring of coal and biomass would immediately reduce the amount of carbon emissions from coal-fired plants by substituting a renewable energy feedstock for some of the coal. Cofiring with biomass could potentially replace at least 8 GW of the U.S. generating capacity by 2010 (DOE, 1997). As an indigenous resource, including biomass in the power generation mix could reduce the risks associated with the fluctuating prices and supplies of fossil fuels (EPRI, 1999). However, successful commercialization will require that the initial capital cost of biopower facilities be reduced, that biopower be integrated into existing power-generation facilities, and that feedstock costs be reduced through coproduction or the use of waste streams as biomass. Every power-generation site can be viewed as a profit center, but introducing new technology into the system will require creativity and economic incentives. As the price for electric power goes down, anticipated profit margins are shrinking and new generating capacity today is dominated by cheap, abundant natural gas

(EIA, 1999b). Therefore, the highly competitive market has reinforced the conservative atmosphere of the power plant industry (Neuhauser, 1999).

The wood pulp and paper industry is a major generator and consumer of electric power. Black liquor combustion is a special case in which a boiler is designed to recover solids for recycling as pulping chemicals. Black liquor recovery boilers (Tomlinson boilers) represent a mature but inefficient technology that has raised safety concerns (Overend, 1999). Cleaner and more efficient black liquor gasification is an area of research that the industry could benefit from and contribute to in proportion to the long-term value of the industry.

R&D alone will not be sufficient for launching new technologies in the market (PCAST, 1997). A comprehensive understanding of the market will also be necessary for the successful adoption of biopower technologies, including a clear understanding of customers' needs and the ability of biopower systems to compete with existing systems in terms of price and performance. So far, the Biopower Program has not effectively promoted biopower systems as viable alternatives to traditional energy sources. Biopower can address the long-term societal needs for mitigating global climate change as well as near-term needs to protect the environment and to make use of industrial biomass wastes. The long-term societal benefits are difficult to quantify, however, and meeting them will require ongoing federal support. However, working with the private sector to capitalize on the near-term opportunities of biopower to reduce environmental effects and address economic concerns will increase the stakeholder base for the commercial use of biopower systems.

"Developing and Promoting Bio-based Products and Bioenergy" (Executive Order No. 13134), issued by the President on August 12, 1999, outlines the importance of developing a comprehensive national strategy for bringing biobased products and bioenergy into the national and international market (OPSWH, 1999a). Two subsequent bills in the U.S. House of Representatives, (H.R. 2819, Biomass Research and Development Act of 1999, and H.R. 2827, National Sustainable Fuels and Chemicals Act of 1999) provide funding to support the initiative and identify the potential value to the rural economy of raising biomass crops as feedstocks for electric power, liquid fuels, and chemicals. Support for the use of renewable energy by federal agencies is included in Section 204 of the Executive Order, "Greening the Government through Efficient Energy Management," which directs that each agency increase the use of renewable energy at its facilities by implementing renewable energy projects and by purchasing electricity from renewable energy sources (OPSWH, 1999b).

With continued investment by government in energy technologies and crops, U.S. farmers could transform a significant portion of our fossil fuel-based economy to a biomass-based economy (Gonzales, 1999). Companies that understand how to take early advantage of a biomass-based economy by successfully selecting and implementing energy options can create a competitive advantage for themselves as valuable as the advantage to the company that developed the technology (Iansiti and West, 1997).

Discussion

Of all the technologies the Biopower Program is investigating, cofiring is the one considered most likely to lead to the near-term integration of research results. As deregulation of the utility industries continues, generation assets in many states are being sold, and the business priorities of the new power companies are changing. A challenge for OPT is retaining the interest of the new owners of coal-fired facilities in the development of cofiring technologies. Cofiring coal with biomass energy crops promises environmental as well as financial benefits to new and old participants in the electric power-generation industry. The Biopower Program's R&D program on modular systems is targeted toward the distributed power-generation industry in the United States and abroad. Engineering design and prototype construction are planned for projects that successfully complete the feasibility stage. Bringing multiple new technologies to market will require a large financial commitment, however, and cost sharing with industry should be in proportion to the value industry would receive from the project.

Based on budget presentations during this study, the committee was not convinced that the Biopower Program has considered the need for long-term partnerships with industry for scale-up, or for determining impacts of such partnerships on the overall program budget. Even though the Biopower Program already has two large demonstration projects (Vermont Gasification and Black Liquor Gasification), they represent only 16 percent of the current budget and are focused on the midterm commercialization of biomass power. Long-term partnerships with industry must be considered to ensure that life-cycle implications of biopower technology can be evaluated.

The Biopower Program's Rural Development Initiative is a unique approach to integrating biomass supplies and biomass conversion for end-use technologies. The Regional Biomass Program has successfully expanded the demonstration of energy crops and provided a link to traditional forest and agricultural communities at the state level. As a partner in the development of a new biomass supply infrastructure, OPT should focus on developing creative mechanisms to encourage businesses in this area.

A common goal of all Biopower Program projects is meeting high environmental standards. Because gasification of biomass resources will be the primary technology for meeting this goal in the long term, the Vermont Gasification Project directly supports this goal. Another project, the Rural Development Initiative, has an outreach program to educate the public about the environmental costs and benefits of biopower projects. The Rural Development Initiative is, in fact, a good example of how multiple objectives can be integrated into a single program. A positive attribute of this project is that a wide range of stakeholders are sharing the financial risk of the development of a unique concept from development to commercialization. A negative aspect is that, because it is such a large component of the Biopower Program and has developed a large cumbersome bureaucracy, the Rural Development Initiative as a whole cannot respond quickly to new issues and must rely on individual projects to respond to the changing environment (OPT, 1999a). OPT management should find ways to protect the fiscal resources of the Rural Development Initiative and minimize the bureaucratic burden on project teams.

For the long term, the development and use of dedicated energy crops is an important element of the overall Biopower Program. Several projects in the Rural Development Initiative are working to introduce energy crops to the agricultural sector. For example, the Bioenergy Feedstock Development Program at Oak Ridge National Laboratory is focused on long-term research on the scale-up of technologies developed by other projects. Introducing new crops to the agricultural community will require a long-term commitment to crop improvement, however, and this will require close cooperation with the USDA.

Findings and Recommendations

Finding. In testimony supporting the passage of H.R. 2819 and H.R. 2827, the value of bioenergy systems to the rural economy and the potential of biomass crops as a feedstock for electric power, liquid fuel, and chemical production were elaborated.

Recommendation. The U.S. Department of Energy (DOE) should consider establishing a center of excellence for bioenergy to bridge internal gaps in the Office of Energy Efficiency and Renewable Energy and create a strategic partnership with the U.S. Department of Agriculture for the development of crops and biobased products. DOE and the national laboratories should assist companies in evaluating, selecting, refining, and integrating bioenergy technologies and opportunities.

Finding. High-quality waste biomass feedstocks offer an immediate opportunity for bringing competitive biopower to the market.

Recommendation. The Biopower Program should highlight the role of waste feedstocks in the current and future biopower market and should leverage existing public benefits for the development and deployment of other renewable energy technologies. This will require an outreach program to engage new participants in the power-generation industry, regional and state program administrators, and local environmental communities.

Finding. The pulp and paper industry is a major user and generator of electric power.

Recommendation. The Office of Power Technologies should consider forming partnerships with the pulp and paper industry to bring cleaner, more efficient black-liquor gasifiers to commercial use. The paper industry should be solicited to commit financial resources to the endeavor proportionate to the long-term value of the technology to the industry.

Finding. The development and commercialization of new technology will require partnerships with research institutions, private industry, and state organizations. Many of these institutions say, however, that the administrative burden and cost of doing business with the U.S. Department of Energy may exceed the value of the funding support. Annual budget uncertainties also make it difficult for a contractor to plan ahead, which ultimately costs the project money.

Recommendation. The U.S. Department of Energy should develop ways of selecting, contracting, and managing projects that reduce the administrative burden on contractors, which takes time and resources away from projects. Multiyear budgeting of projects is one alternative.

Finding. A major economic barrier to the increased use of biomass is the relatively high capital cost of biopower plant construction. Near-term markets are dependent on incentive programs and policies that promote renewable energy.

Recommendation. The U.S. Department of Energy should define the strategic path to achieving the mission of the Biopower Program, including balancing long-term financial commitments to technology and the flexibility to take advantage of opportunities that arise during the transition to a competitive electric power-generation market. As a first step, the federal government should take the lead in adopting advanced biopower technologies to promote commercial acceptance. As a major purchaser of electric power, government sites could provide a baseline market for electricity from biopower and reduce the risk of power plant construction.

Finding. One of the inherent problems of the Biopower Program is its inability to describe its customer base and how biopower could meet its customers' needs.

Recommendation. The Biopower Program should work with relevant components of the U.S. Department of Energy (DOE) Offices of Energy Efficiency and Renewable Energy and the DOE's Office of Science (and coordinate with the U.S. Department of Agriculture and the forest and agriculture industries) to develop a road map for bioenergy research and development activities and to assess its potential value to customers and its customer base.

Finding. During the restructuring process of the electric utility industry, many states have established public benefit research funds for the development and deployment of renewable energy technologies. In many cases, fund administrators have minimal experience working with the U.S. Department of Energy and may not be aware of the Biopower Program.

Recommendation. The Biopower Program should develop a strategy for working with the state public benefit programs to leverage funds and assist in the development of effective initiatives. The Biopower Program could use the Regional Biomass Program more effectively to deploy technology and develop local links for the eventual commercialization of bioenergy crops. The Regional Biomass Program is unique in that it provides a direct connection to state-level initiatives focused on biomass, as well as a link to local environmental communities.

Finding. During the transition to a competitive power-generation industry, many coal-fired power plants are being sold. The rules of the game are changing as the industry evolves. Designing and implementing an effective research program requires understanding the forces driving all participants in the industry.

Recommendation. The Biopower Program should form new alliances with the competitive side of the power-generation market and continue to facilitate the deployment of cofiring coal-fired power plants with biomass from energy crops. The Biopower Program should articulate the financial and environmental benefits of cofiring to the new players in the power-generation market and develop partnerships with them to continue the use of biomass-coal cofiring. Power producers should share in costs proportionally to the value they will receive from the project.

Finding. Long-term opportunities for biomass include energy farming and use as feedstock for the production of a variety of chemicals and other products. New large-scale farming of biomass will affect the agricultural sector, forestry, and land-use policy and may affect biodiversity in areas where energy crops are periodically harvested. Research on genetically modified, efficient biomass crops may facilitate the use of biomass as feedstock but may also raise serious concerns among environmental groups.

Recommendation. The Biopower Program and Bioenergy Initiative should engage the U.S. Department of Energy (DOE) Office of Energy Efficiency and Renewable Energy, DOE's Office of Science, the U.S. Department of Agriculture (USDA), and the environmental community in planning sound long-term research and development programs for promoting the environmentally responsible use of biomass feedstock. The Biopower Program should also take advantage of USDA's tracking of components of the resource base so that early deployment of new biocrops can be monitored to enhance their overall environmental performance.

HYDROGEN RESEARCH PROGRAM

Program Plan and Goals

The Hydrogen Research Program is intended to develop cost-competitive technologies that will improve the quality of the environment and add hydrogen as an energy carrier and energy storage capability to the U.S. energy system. The Matsunaga Act of 1990 (P.L. 101-566) and the Hydrogen Future Act of 1996 (P.L. 104-271) mandated R&D programs that would result in the use of hydrogen for "industrial, residential, transportation, and utility applications." DOE's Hydrogen Research Program, under OPT, has focused on the production, storage, and use of hydrogen, primarily in integrated and distributed fuel-cell systems that can coproduce power, heat, and hydrogen gas. Because DOE recognizes that the technologies and infrastructure for producing and using hydrogen on a commercial scale are probably years away, the hydrogen program has focused on transitional strategies for producing hydrogen from natural gas as a transportation fuel and improving the production of low-cost hydrogen.

OPT's formal strategic plan for the Hydrogen Research Program outlines short-term, midterm, and long-term goals, including the development of a reversible hydrogen fuel cell compatible with other renewable energy systems being developed by OPT. The plan includes R&D on components and subsystems for systems that combine renewable electric power-generation technologies with hydrogen fuel cells.

Program Priorities

DOE has established several programs for hydrogen in response to the Matsunaga Act and Hydrogen Future Act, and some of these programs have overlapping research agendas. OPT's Hydrogen Research Program considers its mission to be the development of cost-competitive hydrogen technologies and systems that will reduce the environmental impacts of energy use and enable the penetration of renewable energy technologies into the U.S. energy mix. The Hydrogen Research Program has adopted the following strategies to achieve its mission (DOE, 1999c):

- Expand the use of hydrogen in the near term by working with industry, including hydrogen producers, to improve efficiency, lower emissions,

ASSESSMENTS OF INDIVIDUAL PROGRAMS

and lower the cost of technologies that produce hydrogen from natural gas for distributed filling stations.
- Work with fuel cell manufacturers to develop hydrogen-based electricity storage and generation systems that will enhance the introduction and penetration of distributed, renewables-based utility systems.
- Coordinate with the U.S. Department of Defense and DOE's OTT to demonstrate safe, cost-effective fueling systems for hydrogen vehicles in urban nonattainment areas and to provide onboard hydrogen storage systems.
- Work with the national laboratories to lower the cost of technologies that produce hydrogen directly from sunlight and water.

The committee believes that these goals are based on the assumption that producing low-cost hydrogen from sunlight and water via renewable sources is many years away unless public policy is changed to focus on mitigating global climate change caused by carbon emissions and on the development of alternatives to fossil fuels. In that case, the development of a "hydrogen-based economy" in the United States would be the logical goal for research.

Status of Research

The Hydrogen Research Program is currently focused on improving the production, storage, and end-use devices for the near term based on the thermal processing of fossil fuels as the source of hydrogen. Longer term renewable energy is focused on the direct production of hydrogen by splitting water into hydrogen and oxygen or by biological or biomass processes.

Hydrogen is the chemist's analog to electricity. Like electricity, hydrogen does not occur naturally in a usable form on earth; it must be generated or produced by consuming fuels or other forms of energy. Like electricity, hydrogen can be used in a variety of applications. Also like electricity, hydrogen is environmentally benign; it can be combusted like ordinary fuels or electrochemically combined with oxygen to generate electricity in fuel cells, forming water (liquid or vapor) as the exhaust product.

Although using hydrogen to produce power would have many benefits, it would also have significant drawbacks and costs. First, energy is required to produce hydrogen and deliver it to the end-user. For example, hydrogen can be produced by fossil, nuclear, biological, or biomass processes or by solar-based electric power-generating systems to produce electricity to dissociate water electrolytically into oxygen and hydrogen. With existing technology, hydrogen requires at least twice as much energy as electricity—twice the tonnage of coal, twice the number of nuclear plants, or twice the field of photovoltaic panels—to perform an equivalent unit of work. Much of this energy is used in the hydrogen production step. For this reason, most hydrogen today is produced from natural gas—an interim solution that wastes 30 percent of the energy in one valuable, but depletable, fuel (natural gas) to obtain 70 percent of the energy in another (hydrogen).

Storage of hydrogen is also a significant barrier because its low density makes it difficult to contain. Cost-effective, practical storage technology will be critical to the commercialization of hydrogen systems. Ever since the unfortunate explosion of the Hindenbergh airship in the 1930s, the public perception (rightly or wrongly) has been that hydrogen is too flammable a gas to use safely. Safety concerns about using hydrogen onboard automotive vehicles or near residential areas are related to the storage, containment, and control of hydrogen gas.

Research Challenges

For hydrogen to achieve its true potential as an energy source, it must be produced and provided to end-users in a clean, safe, transportable medium (i.e., an "energy carrier"). Longer term research for making hydrogen available as a fuel for power generation or transportation is closely related to infrastructure and storage issues. Education to overcome public perceptions of the dangers of hydrogen will also be vital to the success of hydrogen as a fuel.

The first challenge facing the OPT Hydrogen Research Program is to develop better methods of producing hydrogen directly from sustainable energy sources (e.g., biomass, sunlight, etc.) without using electricity as an intermediate step. The second challenge is to develop better methods for storing hydrogen. Like electricity storage in batteries, capacitors, or magnetic coils, hydrogen storage in compressed-gas containers, in metal hydride beds, or as a liquid is bulky, inefficient, costly, and heavy. A major breakthrough in hydrogen energy storage technology would give a major impetus to hydrogen as an energy carrier—just as a breakthrough in battery technology would facilitate the broader use of electricity.

Although the goals for hydrogen use are relatively long term, the Hydrogen Research Program has established a firm foothold in the critical technical areas that can provide incremental improvements:

- Investigations are under way on the direct conversion of biomass to hydrogen, a process that bypasses the electricity step.
- Research is progressing on producing hydrogen from sunlight, photon-catalyzed hydrogen production that does not require electrolysis.
- Low-temperature carbon absorption and advanced magnesium hydride beds are being investigated as potential methods of compact hydrogen storage. Some early projections for hydrogen storage in carbon nanotubes and hydride were extremely optimistic. Subsequent careful review and experimentation showed lower capacity figures. The potential will have to be verified with sound, reproducible science.
- Complementary research is being done on fuel cells in DOE's fossil energy and transportation programs, on coal gasification in the fossil

energy program, and by other fundamental studies in the DOE energy research program.

Commercial Prospects and Barriers

Given the current state of technologies and national priorities, hydrogen is far more likely to be used in the near term as a transportation fuel than as either an energy carrier to replace bulk electric power transmissions and distribution or as an energy storage medium for renewable energy technologies produced in hybrid systems. Low-cost hydrogen from renewable resources should be a research objective of OPT.

Fuel cell technologies will be critical to the use of hydrogen, but costs are still high, and market readiness, especially the lack of infrastructure for both hydrogen-fueled vehicles and stationary power units, is an open question. Advances in production, storage, and conversion technologies will be necessary before hydrogen can be used on a larger, commercial scale.

Timeline for Deployment

The ultimate goal is for hydrogen to replace fossil fuels in every sector of the economy. However, for the next quarter century (and probably longer), low-cost natural gas will be available as the fuel of choice for electric power generation. Unless and until environmental concerns become a high priority, critical issue to the public and politicians, there will be no need for hydrogen as a bulk commodity.

Discussion

In some ways the Hydrogen Research Program is the most intriguing and difficult to assess of all OPT programs. Considering that it was created by congressional legislation and is monitored by a unique statutory body, the Hydrogen Technical Advisory Panel (HTAP), which creates an administrative and management gauntlet, it is a significant achievement that the program works as well as it does. Nevertheless, the tension between short-term and long-term objectives is perhaps even greater for the Hydrogen Research Program than it is for other programs.

Generally, the program plan is well defined and well managed, and the HTAP works closely with the program managers. The research itself appears to be well done, and an organized peer review system is in place. However, in discussions with the committee, even the HTAP panel expressed concerns that too much emphasis is being placed on relatively near-term "technical validation" and the establishment of a distribution infrastructure and not enough on badly needed long-term exploratory and innovative R&D (HTAP, 1998).

Based on materials and presentations to the committee, the Hydrogen Research Program seems to be much more in tune with the overall EERE objectives than many other OPT programs. However, the fundamental short-term problems of hydrogen as an energy carrier (i.e., its low energy density and the lack of a distribution infrastructure) are not addressed in sufficient detail in the program planning documents. For example, high-pressure storage and hydrate storage will both have to overcome significant challenges to become commercially attractive. The limitations of hydrogen storage consequently will require much higher conversion (or total system) efficiency and a substantial direct or indirect policy or regulatory inducement. Therefore, the question arises as to why demonstration projects are included in the current portfolio at all. Because fundamental scientific and/or technological breakthroughs will be necessary for hydrogen to become a viable energy alternative, demonstrations of current technologies will be of limited value. In the words of Richard Rocheleau, Researcher at the University of Hawaii's Natural Energy Institute: "While demonstration of developing technologies can help educate the public, too many resources focused on the demonstration of uneconomical technologies will impede the development of the critical technologies required for the long-term success of the program" (Rocheleau, 1999).

The committee was also concerned about the apparent lack of a systematic process for setting priorities among the production technologies identified in the defined technology pathways. Priorities are particularly important because the program covers a wide field that includes production, storage, and utilization technologies. Regular performance-based reviews would help add focus to the research program.

The Hydrogen Research Program has established successful collaborations with other DOE programs and even outside of DOE. The program should continue to work closely with the Office of Fossil Energy and the Office of Science (formerly the Office of Energy Research) on the development of fuel cells and other hydrogen research, especially if the hydrogen program is eventually obliged to adopt a longer-term focus on hydrogen production technologies and allow others to concentrate on storage and/or end-use technologies. Although the presentation to the committee on new production technologies—fossil-based, biomass-based, and solar/water-based—and their underlying science (Padro, 1999) was interesting, it underscored the committee's conclusion that the most effective contribution of the program is likely to be in the area of hydrogen production.

The notion put forward by the program for using distributed energy systems as part of a distributed fueling pathway is an intriguing way of sidestepping the challenges associated with the lack of a hydrogen distribution infrastructure. This idea blends well with the development of fuel cells for low-emission or zero-emission vehicles.

Findings and Recommendations

Finding. The DOE has a number of programs involving the use of hydrogen, which has created a confusion of effort and responsibility.

Recommendation. The Hydrogen Research Program should be reoriented with a longer term perspective and broader participation by other elements of the U.S. Department of Energy's (DOE's) energy research establishment. The Office of Power Technologies (OPT) should concentrate on research aimed at the production of hydrogen from renewable resources and secondarily on hydrogen storage for distributed power generation. DOE should consider establishing a central point for the coordination of all research on "hydrogen systems," including OPT's hydrogen research and related activities in the DOE's Offices of Transportation Technologies, Fossil Energy, and Science.

Finding. The Hydrogen Research Program does not seem to have a clear methodology for selecting projects.

Recommendation. The Office of Power Technologies should establish a systematic method of setting priorities focused on how resources can best be used. Regular performance-based reviews of projects would improve the efficiency of the program substantially.

Finding. The committee agrees with HTAP's concern that "too much emphasis is placed on relatively near-term 'technical validation' and the establishment of a distribution infrastructure at the expense of badly needed long-term exploratory and innovative R&D."

Recommendation. The Office of Power Technologies should defer its plans for infrastructure development involving hydrogen fueling stations and fuel cells until a practical process for producing hydrogen from renewable resources is in view and a demand for hydrogen begins to emerge.

Finding. Some of the sources and methods for the production of hydrogen that OPT is investigating (i.e., hydrogen for fuel cells or transportation uses) seem better suited to other DOE R&D programs.

Recommendation. The Hydrogen Research Program should focus on the production of hydrogen from all renewable energy resources, including biological methods of production. If the source of hydrogen is natural gas, the program must make a convincing case that the program can produce a superior product for the market. Alternate technologies (including fuel cells) that use natural gas directly should only be used as a reference for setting the goals of the program and should not be the major focus of the program.

HYDROPOWER PROGRAM

The technologies for generating electricity from falling water are among the most mature in the OPT portfolio. Hydropower currently contributes more than 95 percent of the overall renewable energy supply in the United States.[1] Hydropower resources are customarily divided into two general categories: (1) resources that require man-made dam structures with high hydraulic heads (typically 10 to 500 feet) and a power generation output greater than 100 MWe (megawatts electric); and (2) "run-of-river" systems that require minimal dam structures with low hydraulic heads (less than 10 feet) and a power output ranging from a few kilowatts electric (kWe) to 10 MWe. Hydropower can also serve an energy storage function if excess power is used to pump water that can later be used to generate electricity. This is an attractive feature, especially as part of a system that incorporates intermittent renewables.

Plans and Goals

OPT's modest R&D on hydropower is directed primarily at developing more "fish-friendly" turbines for retrofitting existing installations in the United States. About $3.25 million was appropriated in FY99 for hydropower R&D, which represents more than a three-fold increase over appropriations for the last decade (Brookshier and Flynn, 1999). From 1985 through 1998, annual funding for hydropower technology R&D averaged about $1 million per year (Brookshier and Flynn, 1999). The OPT program is leveraged by support from the National Hydropower Research Foundation and cofunding by industry. The program supports the CNES goal focusing on technological solutions for mitigating the environmental impacts of hydropower installations and thus helping to maintain the viability of existing U.S. hydropower resources.

The CNES includes no formal strategic plan for hydropower except for providing enabling environmentally sustainable technologies for existing large hydropower plants. For example, no specific goals have been set for maintaining or expanding the role of hydropower as an option for meeting U.S. objectives of reducing greenhouse gas emissions. No organized reassessment of U.S. hydropower resources is planned as efforts to develop new technology for capturing power from low-head, tidal, or run-of-river resources.

---

[1] Of the 95,000 MW of installed capacity in the United States (77,000 MW conventional, 18,000 MW pumped storage), hydropower currently generates approximately 9 to 10 percent, with about 1,200 plants providing an annual revenue of more than $16 billion (more than 325 billion kWh of electricity).

Modest improvements in efficiency have been realized over the years with evolutionary R&D programs carried out in collaboration with the Tennessee Valley Authority and the Bonneville Power Administration. But for the last 15 years, R&D funding for hydropower has been a minor part of the DOE renewable energy R&D portfolio.

Private sector efforts have also been focused for the most part on environmental issues, primarily fish and water quality problems, associated with hydroelectric resources. Moreover, many private-sector studies are specifically designed to meet regulatory relicensing requirements. These projects are not well coordinated, despite the likely potential for considerable improvements in cost and effectiveness. No comprehensive national plan for engaging the participation of key industry stakeholders or government agencies responsible for hydroelectric resource management has been developed.

Research Priorities

Hydropower is currently the most mature and most developed technology for renewable energy. Hydropower is also an integral part of multipurpose projects for addressing a variety of water management issues (e.g., flood control, irrigation, public water supplies, and recreation), which complicates many of the issues associated with managing hydropower resources. The potential of hydropower to serve as a storage medium for other renewable energy technologies is another possible application of the technology.

Environmental concerns, especially alterations in water quality and adverse effects on fish habitat and aquatic ecology, have led to delays by the Federal Energy Regulatory Commission (FERC) in relicensing existing hydropower facilities, as well as in approving new facilities. Major issues are safe fish passage through existing dam structures and turbomachinery and the maintenance of natural ecological systems. Specific issues include dissolved oxygen levels, minimum stream flows, and land use for farming and recreation. FERC often requires that facilities provide fractional flow bypassing or off-peak operations to maintain sufficient river flow rates to sustain fish migration in connected watersheds without compromising flood control or agricultural irrigation. These requirements have reduced the net generating capacity of relicensed hydropower plants by as much as 8 percent in the present 10-year to 20-year relicensing cycle (PCAST, 1997).

Status of Research

Considerable improvements have been made in the efficiency and durability of large-scale, high-head, hydroturbine and generator equipment. However, making these devices more "fish friendly" and less damaging to aquatic ecology in general will require a better understanding of the causes of fish mortality in turbine machinery. Computational fluid mechanics modeling has identified several viable improvements, including modified turbine blades and other components, aerating turbine designs, and adjustable speed generators. The next steps involve construction and testing of advanced prototype designs to demonstrate both biological and engineering performance to ensure that these technologies are ready for deployment by 2010. Required funding for these demonstrations will be higher than current levels, and some level of government support will be necessary to complete this important phase in the development of commercial-scale units, especially in light of the low cost of fossil fuels and the restructuring of the electric power sector.

Little R&D is being done for smaller scale, low-head hydropower applications, and almost no federal funds are available (Brookshier and Flynn, 1999). Nevertheless, several new concepts have been proposed that could be tested in the field. These first-generation machines, which represent departures from current rotary Kaplan and Francis turbine designs, may sacrifice some conversion efficiency, but they offer substantial ecological improvements in terms of fish mortality, land inundation, oxygen depletion, and silt buildup. Concepts employing horizontal air-foil technology; slow-speed, radial, polymer-composite turbine designs; water-compressed, high-speed, air-driven Francis turbines; power wheels and matrix turbines; and siphon penstocks are all candidates for further evaluation for environmentally sustainable low-head, tidal basin, and run-of-river facilities. These new technologies may greatly expand the potential for hydropower as a renewable energy resource.

Research Issues

In the near term (2000–2010), R&D should be focused on reducing impacts on fish migration to expand and sustain existing generating capacity for high-head and low-head hydropower systems. As our understanding of these impacts improves, FERC licensing and relicensing procedures could be modified to shorten licensing times from the current 6 to 17 years (Mitchnick, 1999). However, the basic R&D infrastructure for both federal and private institutions is currently configured to carry out longer term R&D, particularly for testing improved turbine designs.

In the long term (after 2010), increasing the generating capacity in existing dam/reservoir systems will be the key issue for the Hydropower Program. New technologies will be required to capture run-of-river, low head potential and make more efficient and sustainable use of hydropower resources, with minimal disruption of natural river flows and less degradation of water quality. These new enabling technologies should also substantially reduce or even eliminate silt buildup and flooding.

As with other renewable energy resources, a comprehensive understanding of the resource base can help to maximize hydroelectric resources. For example, coordination of meteorological records with hydropower systems could improve the operational management of existing hydropower installations.

Commercial Prospects and Market Barriers

Some estimates suggest that hydropower generating capacity could be increased by 35,000 to 70,000 MW for the United States with existing dam structures and reservoir systems (Brookshier and Flynn, 1999; PCAST, 1997). However, these increases would require major changes in FERC licensing procedures. These changes would require a much better understanding of the environmental impacts, trade-offs, and uncertainties of hydropower as a basis for quantitative assessments of the long-term sustainability of existing hydropower capacity. Determining these long-term impacts would require the cooperation of many government agencies.

The worldwide potential for hydropower is also very large, and the United States could become a key international supplier of efficient, environmentally sustainable turbomachinery in the growing global hydropower market. However, to date the United States has not been as aggressive as the Europeans in supporting R&D and is likely to lose market share as a hydropower equipment provider. Most U.S. activities overseas are directed toward managing engineering design and construction (A&E firms) rather than on developing new technologies.

The United States could play an important role by providing a better quantitative understanding of long-term environmental impacts of land inundation and river silting associated with large-scale hydropower developments, such as the Three Gorges project in China and the James Bay project in Quebec. This understanding could encourage the use of effective alternative technologies for future international large-scale hydropower projects.

In the United States, numerous environmental concerns are related to hydropower sites (Mitchnick, 1999). In many instances, current public opinion favors removing existing dam structures altogether, or at least substantially diverting water flow around the turbines, to mitigate damage to fish and aquatic ecosystems. On July 1, 1999, for example, the 162-year old Edwards Dam on the Kennebec River in Maine was demolished. Regrettably, alternative designs that are less environmentally damaging are usually not considered as an option.

Timeline for Deployment

The deployment of fish-friendly turbine technology by 2010 is a reasonable goal, provided funds for field testing are available. However, the current Hydropower Program is not aggressive enough to develop new technologies that could expand the hydropower option in the United States, as well as provide more environmentally sustainable hydropower technology to developing countries.

Measures could be taken to expedite the current and licensing process. For example, the latest advances in technical software, such as ISO14000 total life-cycle impact assessment, could help shorten the evaluation time for hydropower development projects (Fisher, 1999).

Discussion

Current R&D on hydropower sponsored by DOE is focused on the development of "fish-friendly" turbine technology. Research has been intent on developing modifications for existing installations and preserving the current 95,000 MW of U.S. capacity. The specific goal of the Hydropower Program is to develop technology that could reduce fish mortality during turbine passage from the current levels of 5 to 10 percent to 2 percent or less (Brookshier and Flynn, 1999).

Although OPT is not currently conducting market analyses or resource assessments for hydropower, earlier estimates indicated that a considerable amount of untapped hydropower was available in the United States. Estimates for new potential range from 30,000 MW from existing dam structures to more than 580,000 MW for all hydrologic resource grades (Mock, 1999; PCAST, 1997). Nevertheless, no organized approach is planned to increase the contribution of hydropower as a part of the 25,000 MW goal of the CNES for 2010. In fact, hydropower is explicitly excluded.

A significant increase in hydropower capacity would require new technologies to make low-head and run-of-river resources economically competitive. For example, ultra-low-head turbines that could be placed in rivers or tidal basins without major water entrapment structures could provide a rational alternative to current designs. Also, siphon penstocks would provide a means of producing power from many earth-filled dams in the United States. Better ways of removing debris automatically from inlet screens could also help maintain productivity.

A coordinated multiagency water management research program should be implemented involving the Bonneville Power Authority, the Tennessee Valley Authority, DOE, the Environmental Protection Agency, the U.S. Army Corps of Engineers, the Bureau of Reclamation, the Fish and Wildlife Service and other government agencies to assess the long-term ecological impacts of existing dams and reservoirs and to develop mitigation strategies to sustain and increase U.S. hydropower capacity. This multiagency program should perform full life-cycle cost-benefit evaluations on the multi-use aspects of hydropower projects, as well as provide quantitative information on the root causes of fish injury and mortality in hydropower machinery.

Findings and Recommendations

Finding. The most urgent need for the preservation of existing hydropower capacity, as well for future development, is for better hydropower conversion

ASSESSMENTS OF INDIVIDUAL PROGRAMS   49

technologies that have higher efficiencies and cause less damage to fish populations. The Hydropower Program's Strategic Plan emphasizes the need for field validations to provide information on the performance benefits of new technologies.

Recommendation. The Office of Power Technologies should accelerate its long-term research on promising, advanced, fish-friendly turbine designs and flow-management schemes, including prototype testing of new and more efficient concepts.

Finding. The CNES has no specific goals for maintaining or expanding hydropower.

Recommendation. Increases in hydropower capacity should be addressed in the U.S. Department of Energy's overall Comprehensive National Energy Strategy.

Finding. Improved quantitative methods should be developed for assessing environmental parameters. These would improve our understanding of the long-term environmental impacts of land inundation, sedimentation and silt buildup, changes in oxygenation levels, and other conditions.

Recommendation. The Office of Power Technologies should focus research on turbine-induced shear and turbulence effects as characterized by computational fluid dynamic modeling, and measurements should be matched with fish damage-inducing mechanisms and acceptable levels of shear stress, flow passage size, rate of pressure reduction, turbomachine length, and number of turbine stages.

Finding. Almost no federal funding has been allocated for the development of low-head hydropower.

Recommendation. The Office of Power Technologies should develop more environmentally sustainable, low-head, hydraulic energy conversion systems for use in run-of-river and tidal basins. The initial focus should be on integrated technology and resource assessment for the United States to quantify the potential of low-head resources. The program should also explore new engineering concepts.

Recommendation. The Office of Power Technologies should develop a coordinated program to assess the benefits of hydropower to meet the storage needs for other renewable energy technologies.

Recommendation. The latest advances should be implemented in technical software, such as ISO 14000 total life-cycle impact assessment methodology, in the evaluation of hydropower development projects. Better evaluations could expedite the current hydropower licensing and relicensing process. If these methods were widely accepted, Federal Energy Regulatory Commission procedures and regulations for relicensing could be reorganized and restructured to reduce lead times substantially.

GEOTHERMAL ENERGY PROGRAM

In 1999, the United States was the largest producer of geothermal electric power with an installed capacity of about 2,800 MW. Worldwide capacity is now almost 8,000 MW with much of the growth in less developed countries (PCAST, 1997). The average growth rate for the last 50 years for geothermal power capacity worldwide has been about 8 percent per annum (Mock, 1999).

The Geothermal Energy Program has historically included a broad range of technologies for tapping the full spectrum of geothermal energy resources for electric power generation. Geothermal resources range from vapor-dominated hydrothermal resources, for which technologies are well known but need refinement to accommodate different environmental conditions, to hot dry rock (HDR) resources, for which the technologies are much less developed and require fundamental R&D to assess their commercial viability. The Geothermal Power Program also supports the development and demonstration of groundwater heat pump technologies, which have shown great promise for cost-effective heating and cooling in most regions of the United States.

Program Plan and Goals

The program budget of $28.5 million for FY99 includes $6.5 million for groundwater heat pump demonstration and deployment projects, which leaves an R&D budget of $22 million for the development of geothermal electric power-producing technologies. Current research is focused almost exclusively on near-term technology, with modest cost-sharing programs with the small, fragile U.S. geothermal industry. The $22.5 million is roughly equally divided among exploration, drilling, reservoir, and energy conversion technology projects. Advanced resource programs (e.g., HDR, geopressured systems, and magma) have been all but eliminated in recent budget cycles.

In the last several years, the program has been working with the U.S. geothermal industry to adjust to lower levels of federal support. At current levels, federal R&D will probably remain focused on near-term opportunities for developing high-grade hydrothermal resources. The earlier significant activities related to developing advanced heat mining approaches for the much larger HDR resource in low permeability formations has been replaced with a program focused on enhanced geothermal systems (EGSs). High-grade EGSs could be located in a variety of geologic settings, including the margins of existing hydrothermal resources and other regions where geothermal temperature gradients are abnormally high.

Research Priorities

Geothermal energy has a large, well distributed resource base. Several grades and forms of geothermal resources in the United States can be used for power generation at duty cycles ranging from baseload to peaking capacity. High-grade, vapor-dominated (steam) and liquid-dominated (hot-water) systems have been developed commercially in the western United States. The largest hydrothermal systems are located in California and Nevada, with the largest of these being The Geysers project (a vapor-dominated system) in northern California. The HDR geothermal resource base in the United States is enormous, with a resource potential of more than 14 million quads (PCAST, 1997).

Like HDR systems, EGSs will also require stimulation to reach commercial production levels. Better injection technologies will be necessary for the efficient use of geothermal resources. EGS resources are located on the margins of hydrothermal resources. Geopressured systems, which consist of hot saline brines under high confining pressures with high concentrations of dissolved methane, are found primarily in the Texas-Louisiana Gulf Coast area with a total resource base of 0.17 million quads (including the energy content of dissolved methane) (Mock, 1999). Finally, magma, with a resource base of 10 million quads, is characterized by very hot regions of molten and near-molten rock associated with volcanic activity.

Even though advanced geothermal resources, such as HDR, geopressured systems, and magma, can use the same surface-based energy conversion equipment and processes as hydrothermal systems, they require advanced technology for deep drilling and reservoir stimulation to lower costs. Most natural hydrothermal, and all HDR and EGSs, operate essentially without gaseous or liquid emissions. Some hydrothermal and geopressured resources may require control technologies to lower natural hydrogen sulfide and other potentially toxic emissions to acceptable levels. Other environmental concerns about the development of geothermal energy include water consumption, subsidence, and seismic risk. Based on current practices worldwide, these environmental issues appear to be either insignificant or controllable. They are not major components of DOE's R&D programs.

Status of Research

Until the early 1980s, U.S. government funding for the development of geothermal technology was sufficient to support a diverse portfolio for short-term, midterm, and long-term programs for hydrothermal, HDR, geopressured, and magma resources. Since then, however, funding has been reduced by a factor of about ten, making it very difficult to sustain a balanced portfolio. Nevertheless, the range of geothermal fluid conditions in which the technology can operate efficiently has increased considerably, and conversion efficiencies of hydrothermal power plants have increased substantially (NRC, 1987).

In a time of declining budgets with few prospects for near-term production, existing field testing programs for less developed technologies that require substantial support for drilling and reservoir stimulation, such as HDR and EGS, have been put on hold or decommissioned altogether.

Despite substantial budget cutbacks in the 1980s for even the best developed technologies in the geothermal portfolio (i.e., hydrothermal technology in high-grade geothermal resource applications), significant progress has been made in drilling technology and down-hole diagnostic methods, reservoir modeling to predict long-term thermal-hydraulic performance, and power conversion methods. In addition, DOE accelerated the development of ground source heat pump technology as a very reliable, cost-effective means of increasing heating and air conditioning efficiency (currently more than 250,000 U.S. houses have been outfitted with heat pumps).

In the long term, the program should focus on advanced concepts, such as high-grade EGSs, and should perhaps build in a demonstration of geothermal systems in lower grade sites (such as in Roosevelt Hot Springs, Utah, or Clear Lake, California) to demonstrate the transferability of reservoir technology concepts. A better understanding of reservoir physical characteristics and behavior is essential to the viability of geothermal power systems.

Commercial Prospects and Market Barriers

Many analysts believe that a substantial fraction of U.S. baseload power could potentially be supplied from a variety of geothermal resources. However, hydrothermal and magma systems are located only in limited areas, mostly along the Pacific coast in the United States, and geopressured resources are located only in the Gulf Coast region. Therefore, geothermal energy could be used extensively in the United States only if the HDR resource can be exploited. DOE could push Congress for specific policy instruments, such as tax incentives, to accelerate the development of geothermal sites and the deployment of geothermal power systems. Improving resource assessment in coordination with the U.S. Geological Survey would also be a significant step towards increasing the use of geothermal resources of all kinds.

The dispatchable characteristic of geothermal electric power supplies has proved to be a virtue for providing both continuous baseload and peak load power. Although the key technology elements for high-grade hydrothermal resources are already in place to enable the program to reach its power capacity growth goals for 2010 and beyond, the economics of resource extraction and power generation are not favorable. The current figures for geothermal electricity from high-grade hydrothermal resources indicate that prices are competitive at 5 to 7 cents per kilowatt-hour (Mock et al., 1997). Unfortunately, natural gas systems are even more competitive in the United States. In fact, no alternative energy system can currently compete with natural gas.

The domestic geothermal industry is currently the world leader in the development of geothermal resources and the installation of geothermal power plants. In the short term, the growth potential for geothermal energy in Asia, South America, and Central America is enormous. However, competition from the Europeans and Japanese, who have been investing more than $80 million per year in R&D on both advanced hydrothermal and HDR technologies is growing. Collaborative projects with Europe and Japan have been limited primarily to informal exchanges of information and conferences.

Timeline for Deployment

Given the size and scope of the Geothermal Program, the committee believes that the strategic goal of the program to increase U.S. capacity to 10,000 MW of electric power and to have 7 million heat pump systems in place by 2010 is not realistic. Although technologies for geothermal power and heat pump systems are available and are capable of meeting this goal, costs are likely to remain too high to encourage substantial deployment, especially in the face of continuing cost and performance improvements in less expensive fossil-fuel alternatives. The goal of increasing geothermal energy capacity internationally in developing countries is even more unrealistic unless substantial policy incentives are put in place to stimulate deployment; incentives could include foreign aid support for joint implementation arrangements under a global environmental agreement.

Despite the relatively small size of the geothermal program's current staff and budget, the Geothermal Energy Program has industriously tried to increase the deployment of geothermal power and heat pump technologies.

Discussion

In light of the significant advantages of geothermal energy as a resource for power generation, it may be undervalued in DOE's renewable energy portfolio. Significant amounts of high-grade resources are available, and geothermal power technologies can operate in a variety of duty cycles (from baseload to peak load conditions) and can be scaled from small ground-source heat pumps in individual homes to several thousand megawatt electric power plants. In addition, the United States has taken the lead in successful commercial demonstrations of geothermal energy for generating electricity and heat at several sites and is the current technology leader in the world among very active competitors in Europe and Japan. With more than $180 million invested in R&D and more than 20 years of experience in the field testing program at the Fenton Hill HDR site operated by the Los

Alamos National Laboratory, many lessons have been learned and a substantial database generated. However, U.S. leadership may be short-lived because the U.S. R&D program is now much smaller than those of overseas competitors.

The strategic plan of the Geothermal Energy Program is well in line with DOE's CNES and has actively pursued the participation of industry in guiding and managing its R&D programs. However, the program does not have a clear tactical plan that includes funding and human resource requirements for achieving the goals in the strategic plan.

Current R&D is focused almost exclusively on the short term and the promise of strong industrial endorsement. The longer term goals of universal heat mining from EGS and HDR resources will require more basic research and are obviously of lower priority to DOE's industry partners, who are struggling for survival in today's low-cost energy markets. Thus, DOE will have to exert strong leadership to balance its R&D portfolio and support longer term objectives.

If drilling costs can be reduced and reservoir productivity levels raised, the long-term prospects for universal heat mining in lower-grade areas would be substantial. This component of research should not be focused on meeting short-term power, on-line objectives. Unfortunately, because of the small size of the current geothermal R&D program, DOE is not currently pursuing these long-term options aggressively enough to make a difference.

The National Advanced Drilling and Excavation Technologies (NADET) Program, initiated with support of the Geothermal Energy Program, has assisted in the leveraged development of advanced drilling technologies to lower costs and open up a larger fraction of the massive U.S. geothermal resource base for competitive power production. Currently, the Geothermal Energy Program provides all of the government funding for NADET. The hope for this initiative was that industry and other government agencies would collaborate and fund general R&D that would support long-term needs in the oil, gas, and geothermal industries, as well as for mining and civil infrastructure.

A critical portion of NADET's portfolio was focused on revolutionary approaches to lowering drilling, mining, and excavation costs and enabling fundamental changes in these industries. Considering the diffuse and scattered nature of R&D on advanced drilling, and despite its potential for reducing costs, improving performance, and lowering environmental impacts, selling the idea to other agencies and to industry has been difficult. On the government side, established programs with small focuses have discouraged the allocation of new funds for such initiatives. On the industry side, a lack of funds, concerns about intellectual property rights, and an unwillingness to collaborate with competitors or government agencies have been limiting factors. The problems with NADET should be carefully analyzed to provide a quantitative basis for moving forward with a new EGS program.

Findings and Recommendations

Finding. The current level of R&D support for geothermal technologies is not sufficient to develop the reservoir engineering science and drilling technologies that would bring down development risks and costs. Therefore, the Geothermal Energy Program cannot effectively pursue a balanced portfolio with short-term and long-term technology objectives.

Recommendation. The Office of Power Technologies must either increase its program funding for the Geothermal Energy Program or make some hard choices about which research it can fund at meaningful levels and cut back or drop the rest.

Finding. Much of the information gained from previous R&D on enhanced geothermal systems has apparently not been used.

Recommendation. The Office of Power Technologies (OPT) should reactivate its programs for the development of advanced concepts for the long term, with its first priority on high-grade enhanced geothermal systems (EGSs) and its second priority on lower grade hot dry rock and geopressured systems. The next steps may involve a commitment by OPT to support one or more field demonstrations of EGS technology. Although several new sites have been proposed for demonstration tests, such as Clear Lake, California, and Roosevelt Hot Springs, Utah, OPT should also consider test sites in lower grade areas to demonstrate the applicability of reservoir concepts to different conditions.

Finding. Improving performance (productivity and lifetime) and lowering development costs will require a better understanding of geothermal reservoir behavior.

Recommendation. The Office of Power Technologies should increase its research and development on reservoir diagnostics and modeling, especially on methods of detecting and enhancing *in situ* permeability.

Finding. Advanced drilling techniques developed in the geothermal program could have wide applicability.

Recommendation. The Office of Power Technologies should attempt to increase the participation of other U.S. Department of Energy offices and other government agencies in research and development on advanced drilling.

Finding. Many nations are engaged in cost-intensive research and development programs to investigate the potential of geothermal energy.

Recommendation. The Office of Power Technologies should increase its collaboration with European countries and Japan on advanced technologies to provide cost-leveraged field testing and enabling reservoir technologies.

Finding. Geothermal heat pumps are a proven technology that could be widely deployed.

Recommendation. As part of its initiative to provide advanced, energy-efficient building systems, the Office of Energy Efficiency and Renewable Energy should focus its efforts on implementing geothermal heat pump technology.

Finding. Better injection technologies will be necessary for the efficient use of geothermal resources.

Recommendation. The Office of Power Technologies should increase its research on the development of injection technologies.

Finding. Much systems work has yet to be performed to improve the efficiency, flexibility, and availability of geothermal power plants.

Recommendation. The Office of Power Technologies should increase its research on power conversion systems to make geothermal power plants more efficient and more flexible so they can function in baseload, load following, and cogeneration hybrid modes.

Finding. Geothermal energy is a widespread but underutilized renewable energy resource. A greater understanding of the widespread availability of geothermal energy will be essential to increasing the use of this resource.

Recommendation. The U.S. Department of Energy should reinstate its resource assessments of geothermal energy at the U.S. Geological Survey and improve coordination among key stakeholders, including the National Renewable Energy Laboratory, the Office of Power Technologies, the Bureau of Land Management, and others.

Finding. Government incentive programs are important to the development and deployment of geothermal-based technologies.

Recommendation. The U.S. Department of Energy should encourage Congress to establish policies that subsidize and accelerate geothermal development. Incentives could include renewable energy portfolio standard requirements, federal tax rebates or loan guarantees for baseload power plant development, home owner tax incentives or rebates for individual ground heat pump systems, and community incentives for the small, remote cogeneration of power.

ASSESSMENTS OF INDIVIDUAL PROGRAMS 57

CONCENTRATING SOLAR POWER PROGRAM

Program Plan and Goals

Solar thermal power plants produce electricity by using mirrors to concentrate sunlight, thus creating the energy to drive a prime mover. OPT describes its Concentrating Solar Power (CSP) Program as a program to develop clean, competitive, and reliable power options with the following goals (DOE, 1999d):

> ... CSP systems are expected to satisfy substantial domestic and international energy needs, contributing 20,000 MW by the year 2020. Consequently, CSP systems are also expected to make a significant contribution to the U.S. effort to reduce carbon emissions in the early part of the 21st century. In response to the changes brought on by utility restructuring and the resulting emphasis on competition, the CSP Program has revised its focus from developing specific technologies to providing technology options to U.S. industry. This effort will enable industry to compete in near-term renewable energy markets and to further reduce costs, allowing for penetration of broader energy markets in the long term.

In response to the committee's questions, the CSP stated that its objectives are consistent with the strategic goals of the CNES: increasing domestic energy production in an environmentally responsible manner; increasing future energy choices; improving global environmental quality; and promoting the development of open, competitive, international energy markets. The CSP Program believes that CSP technologies have the potential to make a significant contribution to OPT's goal of tripling domestic nonhydroelectric renewable energy generating capacity by 2010.

At the time of the committee's review, OPT had no overall strategic plan, although one was under development. The CSP Program, which is also still evolving, has a multiyear plan, developed in 1997 and published in April 1998, that highlights a significant change in direction: "to guide R&D by application factor rather than to complete development on any specific device" (OPT, 1999b). Obviously, the program plan reflects attempts to factor in changes in the market, but no specific goals or metrics have been established to measure progress. Despite these efforts, it will be difficult for the CSP Program to set relevant targets and metrics without an overall OPT plan.

Much of the equipment and many of the technologies in the CSP Program are the same as those used by electric utilities. However, restructuring of the industry has forced the program to change its focus from technologies for central generating stations, such as the power tower, to more flexible approaches, such as power trough systems. In fact, according to the program, "green" markets have become as important as World Bank/Global Environment Facility projects as key drivers for CSP R&D (OPT, 1999b). Therefore, R&D on power-tower systems has been postponed.

Trough Systems

In trough systems, solar energy is concentrated by a field of parabolically curved, trough-shaped reflectors onto a receiver pipe running along the inside of the curved surface. The energy heats oil flowing through the pipe and the heat energy is then used to generate electricity in a conventional steam generator (see Figure 3-1).

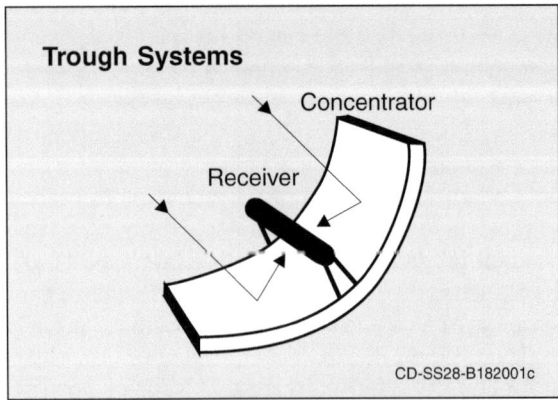

FIGURE 3-1 Solar-trough system. Photo: Warren Gretz. Source: DOE, 1999d.

Dish/Engine Systems

A solar dish/engine system comprises a collector, a receiver, and an engine. Sunlight is collected and concentrated by a dish-shaped surface onto a receiver that absorbs the energy and transfers it to the engine's working fluid. The engine converts the heat to mechanical power, which is then converted to electrical power by an electric generator or alternator (see Figure 3-2).

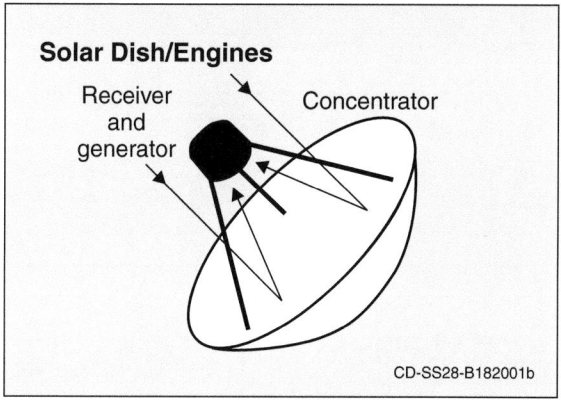

FIGURE 3-2 Dish/engine system. Photo: Warren Gretz. Source: DOE, 1999d.

Power-Tower Systems

In a power-tower system, sunlight is concentrated by a field of mirrors (called heliostats) onto a receiver placed on top of a tower. This energy heats molten salt flowing through the receiver, which in turn is used to produce steam for the generation of electricity. The heat energy retained in the molten salt can be stored for hours or even days before being used to generate electricity (see Figure 3-3).

Program Priorities

CSP's most recent multiyear program plan outlines four subjects for R&D based on their applications rather than the complete development by a set date: distributed power; dispatchable power; advanced components and systems, and alliances and markets. The program goal is to accelerate the commercial readiness of all concentrating solar technologies, advancing the technology in three ways:

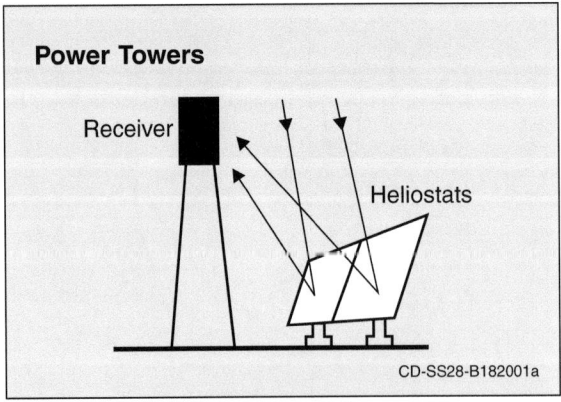

FIGURE 3-3 Power-tower system. Photo: Warren Gretz Source: DOE, 1999d.

ASSESSMENTS OF INDIVIDUAL PROGRAMS

- Develop and demonstrate high-reliability distributed power systems. The primary effort in this area will be improving the reliability of dish/engine systems. Other R&D and field evaluations will be focused on potential new units in the 1–10 kW size range that can serve remote areas, DOE project initiatives in Native American communities, residential sites, and green markets in the United States and other countries.
- Reduce costs of dispatchable solar power. The principal focus in this area will be on advances in solar trough systems (e.g., concentrators, better storage technologies, and advanced operations and maintenance methods) to reduce costs per kWh and a second-generation trough system. Future R&D will include solar/natural gas hybrid and high-temperature systems.
- Develop advanced components and systems. R&D in this area will focus on improving mirrors, subsystem engineering, heat pipes, converter evaluations, and other components to bring costs down and improve reliability. These advancements will broaden the market application for CSP systems, particularly in the domestic arena.

To complement R&D, a fourth phase in the CSP plan will focus on strategic alliances. The goal of this phase is to assist program managers in maintaining up-to-date information on potential markets and industry goals and to make the technical expertise of the national laboratories available to U.S. corporations and government agencies. This program area includes application studies and systems analyses, updates of technology road maps, project feasibility assessments, and the formation of a CSP advisory council. On the international front, collaboration will be channeled through the International Energy Agency's Solar Power and Chemical Energy Systems (IEA SolarPACES) agreement.

Like many other OPT programs, CSP's priorities have been developed through a combination of "bottom-up" and "top-down" exercises. For special projects, funding typically is allocated from the top down. A bottoms-up, iterative approach is used to balance OPT demonstrations of economic viability of a technology via early deployments with longer range R&D by the national laboratories. As a result, the program's management approach is neither strategic nor coordinated with the plans and goals of other programs; CSP's portfolio is mostly politically driven; and no hard measures have been established for measuring progress or allocating funding.

Status of Research

The CSP Program relies heavily on the technical expertise of the national laboratories (specifically, Sandia National Laboratories and the National Renewable Energy Laboratory, which are referred to collectively as SunLab). CSP managers believe the level of scientific excellence at both laboratories is without peer in several areas of solar research (e.g., heat-pipe receivers and structural facets). Because CSP technologies are based on common construction materials, and program content is slanted towards engineering, system testing, and low-cost processing of system components, CSP believes there is no need to coordinate its R&D with science-focused organizations, such as DOE's Office of Science, the NSF, or the National Aeronautics and Space Administration.

Research Issues

Although CSP has a less than urgent need for cutting-edge science, CSP program managers cite a need for continued improvements in solar reflectors. Mirrors are a substantial component of CSP system costs, and the current technology is considered marginally adequate. An ideal mirror would be highly reflective, self-supporting, weather resistant, soil resistant, nearly maintenance free, lightweight, unaffected by wind loads, nondegrading for decades, and cheap to manufacture. CSP managers say they have already tested numerous designs and will continue to evaluate new designs. Resource assessments, they believe, are the key to identifying cost-effective applications of CSP technologies; they also create market opportunities by making potential users aware of the magnitude of the renewable resource that is available to them.

Responses to the committee's questions indicated that the CSP Program faces no serious R&D issues. The responses also suggested that coordinating or leveraging more fundamental R&D being done in other offices and agencies was not necessary. The committee feels that this attitude reflects a very narrow view of the challenges facing CSP technologies. For example, better engineered components and subsystems may be necessary for the demonstration of near-term CSP technologies and may require fundamental progress in materials for high-temperature and heavy duty-cycle components. Ensuring the reliability, and thus the economic attractiveness, of CSP systems will probably require progress in R&D being done elsewhere.

Commercialization Prospects and Market Barriers

Overall, the commercial prospects for CSP technologies are not very promising. Despite claims by the CSP Program that the cost of power-trough technology is presently about 11 or 12 cents per kilowatt-hour, industry analysts suggest that only 16 cents per kilowatt-hour has been demonstrated. DOE's projections indicate that the cost will be about 8 cents per kilowatt-hour in the next three to four years (DOE, 1999e). However, DOE's projections have been notoriously optimistic in the past because DOE management has had to show near-term commercial viability to secure funding. CSP technologies are especially vulnerable to overly optimistic promises because, although the technologies have compelling potential, significant deployment is still years away.

During this study, the Arizona Public Service (APS), a company with a strong interest in dish technology, made a presentation before the committee. APS recognizes that participating in technology development would give the company a competitive advantage in the marketplace. Therefore, APS has entered into a partnership with Science Applications International Corporation (SAIC), one of the few dish technology companies still active, with the goal of demonstrating the viability of the technology. If in the future the state of Arizona establishes a solar renewable portfolio standard (i.e., a requirement that a certain percentage of total electricity be generated from solar energy), APS hopes to meet that requirement with the lowest cost, modular, solar technology available. APS believes that dish systems have the potential to meet that need, but the technology is in the very early stages of development, which is reflected in current costs. Therefore, APS is also exploring photovoltaics technology, which is better positioned to meet this market demand. From the committee's perspective, the interest by APS in dish technology indicates that an early niche market is available.

No private market has been identified for power-tower or solar-trough technologies. Architecture and engineering and engineering construction firms, such as Bechtel (which participated in the Solar 1, Solar 2, and Solar Electric Generating System [SEGS] plants) and CH2M HILL (which participated in the SEGS solar-trough plants), are the most likely companies to bid on jobs for engineering design and construction, but these projects will require government financing. At present, unless there is a significant market intervention by the federal and/or state governments, an economically feasible project in the United States will not be possible.

Project viability may be more likely in foreign markets, but those projects would also require significant intervention by a financial institution. Under current conditions, the most likely sources of financial support are the World Bank or intergovernmental agencies, such as U.S. Aid for International Development (USAID). U.S. companies, however, would then be at a disadvantage because host nations will want to derive the economic benefits of construction and operation locally. Therefore, U.S. supported financing will be necessary for U.S. stakeholders to benefit.

Timeline for Deployment

CSP's response to the committee's questions about the viability of the CSP technologies in the marketplace seem to indicate the readiness for deployment of some technologies (OPT, 1999b):

> Solar trough generating facilities (totaling 354 MW) have been operating in domestic applications for a number of years. DOE sponsored R&D has resulted in improved reliability, reduced operations and maintenance costs, and higher electricity output, to the point where these facilities are now operating more inexpensively and with higher reliability than when they were first built (1980s).

> The 354 MW of installed solar troughs, built in the 1980s, remain economically viable today without external subsidies. However, they were erected at a time when investment tax credits were available, from both the Federal government and the State of California, to help offset the capital cost.

The solar troughs built in the 1980s are solar/gas hybrid systems that can produce dispatchable energy (i.e., energy produced on demand for an electric power system grid). A new solar-trough plant constructed today incorporating more advanced technology would have lower energy costs by several cents per kWh but would still be economically viable in very few domestic market areas. The best current opportunities for near-term systems are in subsidized international procurements for village power and grid connected applications, such as those under development by the World Bank or the United Nations' Global Environment Facility (GEF) projects.

Power-tower technology is not competitive yet either. This technology has been demonstrated at the Solar 2 Facility in Mojave, California, but no utility or other company has proposed building commercial plants.

Dish/engine systems are not yet commercially viable. The CSP Program hopes the recent decision to focus on domestic applications (rather than GEF/foreign projects) and the emphasis on a 10 kW unit will facilitate the market entry of this technology by 2004. CSP projects that initial sales will be to Native American organizations (which are often subsidized) and to domestic and foreign remote users (who expect to pay higher prices for electricity). Further improvements will lower costs and may permit penetration into the distributed-power markets, beginning perhaps about 2007 (OPT, 1999b).

Discussion

Although CSP technologies promise compelling benefits, as well as low costs, no specific goals or objectives have been established to determine an appropriate level of federal investment. The problems encountered by the CSP Program are similar to the problems facing all OPT programs. OPT has not been able to chart a strategic course for the development of technologies that have both strong positive and negative features. For example, ease of hybridization and cost effectiveness of solar towers, dishes, and troughs (superior to photovoltaics) in the near future are offset by the high capital cost of entry (for central station power towers), the immaturity of solar-dish technology, and the lack of interest in the private sector, even in a proven technology, such as the solar-trough system. This dilemma raises questions about the goal of federal R&D programs in general. Is it appropriate for a federal program to preserve a technology option that requires a significant investment? If the initial market is overseas, should a U.S. government program be addressing it? What should the short-term and long-term commitment of U.S. industry be to this technology?

CSP technologies seem to be relatively well developed for solar thermal applications. Although some technological improvements will still be necessary, for the most part engineering (e.g., electronic controls, new drive concepts, lighter and lower cost mirror facets, etc.) will be necessary rather than basic research. Although more R&D will be necessary to develop a "manufacturable" Stirling engine, this work does not have to be done by OPT or by the national laboratories.

Most of the current interest in CSP for central generating stations is overseas. Small, village power or distributed power-generation schemes appear to be the only possible domestic applications, and these would have to be hybrid systems because they can only generate power in daylight. If the OPT concludes that this is a viable market for its CSP technologies, then research should be concentrated on meeting these needs. Despite this possibility, however, the committee believes that the international markets for CSP technologies is limited and that only small, incremental improvements are likely to result from continued R&D.

The arguments for continued research in this area (i.e., to maintain solar thermal technology as a future option) are not very compelling because the technology is already essentially deployable (i.e., the likelihood of major breakthroughs that will affect cost and performance is small and/or not commensurate with the potential payoff). The absence of buyers for a U.S. solar thermal facility speaks for itself, and there is no reason to expect the situation to change in the next 10 to 20 years.

Collaboration between program managers at the national laboratories and DOE headquarters seems to be more effective in the CSP Program than in other OPT Programs. However, the CSP Program does not have broad private sector input. The private sector stakeholders that are involved are supported by (and dependent on) government contracts and, therefore, may not provide an objective picture of market barriers to commercialization. In fact, the CSP program seems to be at a crossroads. Although CSP technologies promise to provide clean energy, the issues of cost and siting must be addressed before the benefits can be realized.

Findings and Recommendations

Finding. For all intents and purposes, power-tower and power-trough technologies could be deployed today. However, no buyers have come forward for initiating commercial operations in the United States.

Recommendation. The Office of Power Technologies should limit or halt its research and development on power-tower and power-trough technologies because further refinements would not lead to deployment.

Finding. Solar dish/engines seem destined for niche operations and likely to be used in hybrid systems with other power-generation technologies in remote off-grid areas.

Recommendation. The Office of Power Technologies should reassess the market prospects for the solar dish/engine technologies to determine whether continued research and development would result in a technology that warrants further expenditures.

SOLAR PHOTOVOLTAICS PROGRAM

Photovoltaic systems require only sunlight to produce electricity, produce no effluents, and have only one direct impact on the environment—the space taken up by the arrays. Two critical problems have prevented the widespread production of electricity by photovoltaic systems. First, the cost of manufacturing photovoltaic modules is relatively high. Second, a means of supplying power when the sun goes down must be used (e.g., stored energy or another source of power, such as a diesel generator).

Program Plan and Goals

The stated mission of the DOE's Solar Photovoltaics Program is to implement a balanced, aggressive R&D program to develop clean, competitive, reliable solar photovoltaic power technologies for the twenty-first century. Table 3-1 shows the long-terms goals in terms of the target price of electricity and other factors. Manufacturing costs are expressed in dollars per watt of power produced; module efficiency (i.e., how much captured light energy is actually converted to electricity) is a significant factor in determining module cost.

Program Priorities

The DOE touts photovoltaic technology as a way to generate clean, affordable energy in the years to come. In fact, photovoltaic systems can help reduce

TABLE 3-1 Long-Term Goals for Photovoltaic Technologies

	1991	1995	2000	2010–2030
Electricity price (¢/kWh)	40–75	25–50	12–20	< 6
Module efficiency[a] (%)	5–14	7–17	10–20	15–25
System cost ($/W)	10–20	7–15	3–7	1–1.5
System lifetime (years)	5–10	10–20	> 20	> 30
U.S. cumulative sales (MW)	75	175	400–600	> 10,000

Source: DOE, 1999f.

[a] For commercial flat-plate and concentrator technologies.

greenhouse gases produced in the generation of electric power, provided manufacturing costs can be brought down by R&D and economies of scale. Bringing down costs will require steady, profitable growth in the photovoltaic business, which could be supported by relevant R&D. Every R&D project undertaken by DOE buys down the monetary risk of private sector R&D. But because all of OPT's R&D programs compete for the same funding, cost sharing with industry is a major aspect of the photovoltaics program.

Current worldwide sales are about 200 MW per year, with shipments divided about equally among the United States, Japan, Europe, and other markets. If total system costs (modules, array structure, and power conditioning equipment) are included, the market value is about $2 billion. More than 90 percent of all photovoltaic modules sold today are made up of single 1-watt silicon cells connected to form a module of 40 to 100 cells. If a single module cannot supply the power needs of a specific application, the modules are connected to form an array.

The restructuring of the utility industry has created long-term opportunities and short-term difficulties for the photovoltaic program. The primary opportunity is that distributed power-generation technologies (such as photovoltaic systems) are likely to play a role in power generation commensurate with their technical and economic capabilities. Short-term difficulties include uncertainties about who will own the power-generation sources and distribution channels and who will set the standards for interconnecting distributed power-generation sources. In the past, the photovoltaic program supported the establishment of standards for interconnecting photovoltaic systems into the electrical grid, and these standards are now being used as a basis for all distributed power-generation sources.

Electricity generated by photovoltaic systems now costs about 30 to 40 cents per kWh, which is economical in locations that do not have easy access to the electric grid, such as roadside emergency phones, highway signs, navigational buoys, electric fences, and remote sensing stations that require small amounts of power, usually 100 W or less. Photovoltaic systems are also being used to generate power for water pumps and off-grid homes, which usually require 500 to 5,000 W. These applications indicate that photovoltaic systems can be competitive with other alternative technologies, such as diesel generators.

Status of Research

Crystalline silicon, the most advanced photovoltaic technology, is in the deployment stage. About 96 percent of the current world market is based on crystalline silicon technology. Several U.S. manufacturing facilities have the capacity to produce more than 20 MW per year of silicon modules. The needs of the remaining 4 percent of the world market are being met by thin-film photovoltaic technology, the most advanced of which is amorphous silicon. As thin-film costs decline, sales are expected to grow at an annual rate of 15 percent to 25 percent per year (Birkmire, 1999).

Several small facilities for producing amorphous silicon are working on meeting design production capacity, and a limited number of low-efficiency modules are available for sale. The manufacture of thin-film modules of cadmium telluride and copper indium diselenide is at the pilot-plant level of production. A small number of modules of copper indium diselenide and cadmium telluride are available for sale. A laboratory-scale program to develop very thin silicon (under 10 microns) modules is under way, but only very small devices have been manufactured so far.

Research Issues

The photovoltaic program recognizes that a substantial part of the life-cycle cost and the reliability of photovoltaic systems depends on the "balance of systems" components and integration. Balance of systems components refers to all elements of the system except the photovoltaic module. Because photovoltaic systems use batteries and power electronics differently from most other systems (i.e., deep discharge and recycling time), specific test and evaluation procedures have been developed for storage systems. Some technical challenges associated with storage or hybrid generation systems (i.e., combining photovoltaic with other types of power generation) must still be overcome.

The research needs in photovoltaic systems design really boil down to a need for sound, creative engineering. If practical, "manufacturable" systems are established as the goal of the research process, research decisions will be based on practical goals and realistic priorities.

Photovoltaics are a unique solar-power technology because the solar cells are sensitive to the spectral content of the solar resource, as well as to the total incident solar radiation. Therefore, evaluating the quality of the solar resource (e.g., solar insolation, spectral content, daily and seasonal variability) is an important economic criterion for siting photovoltaic systems. The Solar Photovoltaics Program has supported resource assessment for many years, including the development and validation of solar radiation models and the development of solar resource databases.

A critical outstanding issue for the commercial-scale manufacture of thin-film modules is the lack of a capability for design, operation, and control of deposition units and other steps required to make modules continuously. All layers of thin-film modules will have to be deposited on substrates moving at a rate of 1 to 10 meters per minute to reach cost reduction goals. This will require the development of effective *in situ* measurement and model-based controls, which will require innovative laboratory experiments and creative engineering. To implement model-based control schemes, an analysis of the reactions and the reactor for the deposition process will be necessary, and experimentally verified mathematical models will have to be developed. This fundamental chemical engineering analysis will also be essential for the design, operation, and control of continuous processes for producing thin-film modules.

As the restructuring of the electricity industry proceeds, state programs are expected to undertake R&D on photovoltaics. In many states, SBC funds have been set aside for the purpose of developing renewable energy portfolios. These programs represent an opportunity for OPT's photovoltaics program.

Commercial Prospects and Market Barriers

The primary technical barriers to the deployment of photovoltaic systems are cost and performance. The cost is currently at least three times that of electricity from conventional sources and up-front costs are relatively high. Photovoltaics are an intermittent power source. Therefore, to provide power on a demand basis, photovoltaic systems must be combined with other generation sources (e.g., diesel generators and/or batteries).

The photovoltaic program is attempting to facilitate deployment by reducing these technical barriers. For example, work is being done on the development of consensus standards and codes to increase product acceptance; work is also being focused on improving component characterization procedures and developing reliability tests. The Solar Photovoltaics Program also supports the development of new products to address the needs of specific market segments. Photovoltaic systems are currently the lowest cost option for electrification in many developing parts of the world, a very large market for U.S.-based photovoltaic products. In fact, more than two-thirds of U.S. photovoltaic products are exported.

Manufacturing costs have been reduced by a factor of 10 in the past decade by a combination of efforts by industry, universities, and national laboratories, partly coordinated by OPT's photovoltaic program. Although exact figures for manufacturing cost were impossible to obtain because manufacturers consider them to be proprietary information, available data suggests that the current cost of producing single or polycrystalline silicon modules is between $3 and $4 per watt. Therefore, it is debatable whether businesses are profitable, but the committee believes that some manufacturers have made small profits in the past couple of years.

Manufacturing facilities for the largest firms producing silicon-based cells and modules currently produce about 20 to 30 MW. Total annual worldwide manufacturing capability is 150 to 200 MW, although this figure is subject to the same level of error as the estimates of manufacturing cost. Because the manufacturing process for silicon cells is modular, there is little economy of scale in the manufacture of such modules. The modules comprise single cells obtained from some type of crystal growth operation. Therefore, an increase in production capability would require another crystal growth unit, as well as equipment for sorting cells, assembling modules, and encapsulating the module. Doubling the capacity requires doubling the number of units.

The photovoltaic community generally agrees that significantly lower manufacturing costs can only be achieved for thin-film modules. Thin-film semiconductors are only a few microns thick (compared to the several hundred microns in silicon cells), and they can be deposited continuously on a moving substrate. Four thin-film materials are in various stages of development.

Projects in the photovoltaic program, such as the Photovoltaic Manufacturing Technology (PVMaT) Project, have focused on reducing manufacturing costs for silicon cell modules with well designed cost-shared contracts. More "fundamental" research at the National Renewable Energy Laboratory and universities has led to significant improvements in the performance of small cell devices and a steady increase in conversion efficiencies. Efforts to encourage cooperative discussions of critical technical issues affecting thin-film technologies have had mixed results and have not been nearly as effective as a consortium program supported by the Defense Advanced Research Projects Agency.

In general, as issues of intellectual property and licensing arise during the development of technology, knowledge is not always shared in consortium-based research. Knowledge originally intended to remain in the public domain may become the subject of intellectual property disputes as consortium members sort out public and private contributions to technology development. These disputes can inhibit the deployment of technology. Therefore, the photovoltaic program should set practical goals and realistic priorities to avoid these problems.

Timeline for Deployment

Photovoltaic markets worldwide have been growing at an average rate of 20 percent per year over the past 10 years (Birkmire, 1999). In 1998, worldwide shipments of photovoltaic modules totaled nearly 152 MW, with U.S. companies supplying 54 MW of this total. An OPT road map to meet photovoltaic goals is under development.

Discussion

OPT's Solar Photovoltaics Program is subject to political pressures from outside and inside DOE; nevertheless, the program has been well managed and responsive to industry pressures, some of which have been well formulated and some of which have been misguided. Effective progress in developing low-power, off-grid applications has kept many firms in business and is partly responsible for today's billion-dollar industry.

Despite the promise and potential of solar photovoltaic technologies, however, DOE and the national laboratories should refrain from raising unrealistic expectations about the role of photovoltaic systems in supplying electric power. Researchers in renewable energy should avoid the temptation of proclaiming "great technology breakthroughs," which can divert attention from basic long-range materials development, raise unrealizable expectations, and cause unnecessary political and public relations problems for all renewable technologies.

Findings and Recommendations

Finding. Reducing life-cycle costs and improving the reliability of photovoltaic systems will require greatly improving the "balance of systems components" for power conditioners and storage devices.

Recommendation. The Solar Photovoltaics Program should focus more on balance of systems components, which will require expanding and refocusing basic research on eventual commercial-scale manufacture. To facilitate the setting of priorities and effective budgeting, the Office of Power Technologies should ensure that more systems analysis is done.

Finding. Currently, "research needs," particularly in photovoltaic system design, really require sound and creative engineering rather than more research.

Recommendation. The quality of research and engineering will have to be improved dramatically for the next generation of photovoltaic devices. The Solar Photovoltaics Program should distinguish projects that require original scientific research from projects that require creative, competent engineering. In addition, fundamental and "applied" research will have to be better integrated for effective planning and for setting priorities.

Finding. Improvements in energy storage by 2020 will be necessary for the widespread deployment of photovoltaics systems.

Recommendation. The Office of Power Technologies should focus on the development of storage technologies that will complement photovoltaics operations, including batteries with deep discharge and recycle characteristics.

Finding. The promise of inexpensive solar-electric power generation has not yet been realized.

Recommendation. The Solar Photovoltaics Program should give top priority to the development of sound manufacturing technologies for thin-film modules. Much more attention should be paid to moving the technology from the laboratory through integrated pilot-scale experiments to commercial-scale design. This will require much more engineering expertise than has been utilized to date.

Finding. If researchers were made aware early on that eventual manufacture is critical to the success of their research, laboratory-scale experiments in the "fundamental research effort" would be greatly improved.

Recommendation. The Solar Photovoltaics Program should focus its efforts on the end goal (i.e., the manufacture of photovoltaic systems). Most laboratory-scale experiments could, with very slight modifications, provide critical information for eventual commercial-scale design. The program should make a concerted effort to integrate fundamental research and basic engineering research.

Finding. State governments and agencies are involved in R&D and will have the opportunity to promote the deployment of solar photovoltaic systems.

Recommendation. The Solar Photovoltaics Program should develop a mechanism for interacting with state programs that encourages the use of photovoltaic technology.

Finding. The potential contribution of solar photovoltaics to meeting national energy requirements has been recognized overseas.

Recommendation. The Solar Photovoltaics Program should focus more on meeting the needs of international markets, which is where most photovoltaics technology is being sold.

Finding. The Solar Photovoltaics Program has placed too much emphasis on improvements in efficiency as the sole indicator of R&D progress.

Recommendation. The Solar Photovoltaics Program should reevaluate its metrics for determining progress. Efficiency should not be the only measure of progress.

Finding. Many unrealistic promises have been made over the years about the potential of solar-electric power generation. Unrealistic promises undermine the credibility of the Office of Power Technologies.

Recommendation. The Office of Power Technologies should institute a process for regular peer reviews of the photovoltaics program to determine directions for future research.

Finding. The photovoltaics community has principally focused on cell conversion efficiency and hardly at all on other areas, such as manufacturability.

Recommendation. The Solar Photovoltaics Program should modify its fundamental research program to ensure that researchers recognize and focus on the need for eventual commercial-scale production. "Great technology breakthroughs" with limited applicability are of little if any use. Projects should be focused on providing a basis for long-range materials development for photovoltaic arrays.

WIND ENERGY PROGRAM

The environment for the commercial development of wind power technology has changed dramatically in terms of technological maturity, the evolution of domestic and international markets, and public policy. Like most other renewable energy technologies, wind technology has endured the roller coaster ride of large then greatly diminished federal investment in R&D and public policy that provided substantial incentives for commercial deployment and then eliminated them. As a result, at least in the United States, the industry has been greatly diminished, in terms of both technology suppliers and resource developers. Although the overall capacity for developing wind technology and deploying it widely in domestic and overseas markets remains strong, sustained federal support will be required for the next decade.

Program Plan and Goals

The DOE Wind Energy Program is organized around the following three functional areas (Thresher and Hock, 1999):

- applied research, designed to provide the fundamental underpinnings of the program, including design codes and standards, techniques, and databases for various segments of the wind industry
- turbine research, which is focused on advancing conceptual and engineering designs of wind turbines and subsystems primarily by industry/laboratory partnerships; initial performance is verified through experience with small numbers of turbines
- cooperative research and testing, which addresses near-term problems and the needs of industry by establishing certification testing and procedures, developing international standards, tracking industry performance, and sponsoring analysis and testing at the National Wind Technology Center at the National Renewable Energy Laboratory

In addition to these principal program areas, in June 1999 Secretary Richardson announced a DOE initiative, Wind Powering America, to increase the use of wind energy in the United States (DOE, 1999g). The goals of the initiative are: (1) to provide at least 5 percent of the nation's electricity by wind generation by the year 2020, to have installed more than 5,000 MW by 2005, and to have installed more than 10,000 MW by 2010; (2) to double the number of states with more than 20 MW to 16 by 2005 and to triple the number to 24 by 2010; and (3) to increase the contribution of wind power to federal electricity use to 5 percent by 2010.

The overall purpose of Wind Powering America is to accelerate the commercial adoption of wind technology in U.S. electricity production. The Wind Powering America program is neither managed nor funded by OPT's Wind Energy Program. If Congress appropriates the funding for this initiative, it will focus on technology transfer through regional partnerships, an increase in the use of wind power at federal facilities, and further technology development. The initiative will attempt to achieve these goals by building on public and private sector efforts to support the development of wind power. The action plan includes the following elements:

- Build an awareness of wind's benefits to build consumer demand.
- Increase federal wind use to promote technological maturity.
- Foster appropriate policy choices to provide a supportive investment environment.
- Support a lowering of barriers to overcome institutional biases.
- Advance U.S. technology to ensure competitiveness in the global marketplace.
- Support the development and use of small wind technology.
- Communicate successes to maintain momentum.
- Educate the American public to promote environmental consciousness.
- Integrate wind into other federal programs to increase the breadth of its constituency.

Status of Research

Wind technology has improved substantially in the last two decades. The most important impact of these improvements has been to lower costs, which have fallen from more than $1.00 per kWh in the early 1980s to 5 to 6 cents per kWh today. In the most favorable wind regimes and with state-of-the-art technology, costs are below 5 cents per kWh (Thresher and Hock, 1999). These dramatic improvements are largely the result of substantial improvements in wind turbine designs (15–30 percent improvement in energy performance since the 1980s); much more efficient and less costly power electronics (contributing energy gains of up to 20 percent); improvements in materials performance; improvements in construction methods; and a vastly improved understanding of wind patterns and siting.

These advances were crucial to the wave of commercial development in the mid-1980s, but more improvements will be necessary to bring costs down for large-scale deployment in the United States (OTA, 1995). Current technology will probably be commercially viable in many overseas markets, although additional advances in power electronics would be helpful. New materials for advanced wind turbine design could improve performance over a long period of time.

A perennial challenge to high efficiency is the generation of electricity under widely varying wind conditions. With recent advances in power electronics, variable frequency power can be converted into a constant voltage and frequency. Most current wind turbines operate at a fixed rate of rotor rotation to synchronize with the power grid, which limits generation at low wind speeds and limits the range of winds in which the turbine can operate. Advanced power electronics converters have yielded much higher efficiencies at low wind speeds and extended the range of operations and the range of locations where wind resources can be commercially developed.

Research Issues

Reducing capital cost is the principal driver for R&D in wind technology, and continued investment will be necessary for the United States to remain competitive in short-term emerging markets, especially in developing countries, and in long-term domestic markets. Great strides have been made, such as improvements in the durability and reliability of turbines, but wind turbine technology is still far from mature, and substantial gains could be made with relatively modest investments. Nevertheless, the challenges for this new technology in an environment of low energy prices are daunting.

Continued research will be necessary to improve performance in turbulent flow. Advanced computational modeling to identify potential improvements in performance in various wind regimes will help prolong turbine life and reduce cost. Development of lightweight structures that can passively reduce loading and extend the fatigue life of turbine blades and other components should be a high priority. Improvements in manufacturing technology could reduce costs, and the development of direct-drive, variable-speed systems is likely to be the key to major cost reductions. The development of advanced controls and improved gearboxes appear to be well within the capabilities of industry.

Many of the environmental concerns associated with wind power identified in the 1980s have been addressed. A better understanding of bird migration patterns and pathways, as well as changes in support structures, like tubular towers that discourage birds from landing, have addressed most concerns about bird deaths. Improvements in technology have also reduced noise levels to generally acceptable levels up to within several hundred yards of modern wind turbines. Concerns about aesthetics and land use are still outstanding, but overall environmental issues are no longer major impediments to the commercial viability of wind power.

Siting and resource assessment will be crucial for successful commercial deployment. Resource assessment has long been a generally neglected component of the R&D program. Providing industry with the ability to make detailed characterizations of wind resources will enable industry to improve the commercial viability of proposed wind projects.

The emphasis on research, testing, and field verification should be renewed. According to OPT's strategic plan, the mission of the Wind Energy Program is to complete the research, testing, and field verification necessary for development of fully advanced wind energy technologies that will lead the world in cost effectiveness and reliability.

Coupling wind farms with effective energy storage will dramatically increase the value of wind power as a power generating resource. Because wind is intermittent, wind power will have difficulty competing with technologies and resources that have much higher capacity utilization; cost-effective energy storage would help enormously.

Commercial Prospects and Market Barriers

In the last decade, the development of wind energy resources for power production has become a global market, and many countries, through both industry and government enterprises, are investing heavily in R&D. OPT's Wind Energy Program, combined with temporary substantial federal and state renewable energy subsidies, have been responsible for the U.S. lead in technology development.

The technology pipeline for wind energy in the United States, like the technology pipeline for most other renewable energy sources, is in transition from government-dominated R&D to R&D focused on meeting market demands. However, because of current low energy prices in the United States and the elimination of the Public Utility Regulatory Policies Act of 1978 and other federal and state incentives for the development of renewable energy technologies, the market has been limited. Hence, despite the United States having some of the most favorable wind regimes in the world, the principal emerging markets have been overseas.

Today, many states are establishing funds accumulated from the restructuring of the electric utility industry to support the development and deployment of alternative energy technologies. Because wind technology is one of the most commercially viable alternatives and because the United States has many favorable locations for wind technology, the benefits of wind technology in the next decade will probably be sizable. In 1999, the U.S. market, fueled by these state programs, was reborn. More than 1,075 MW of new wind capacity was installed in the last year, and wind turbine capacity in the United States now exceeds 2,500 MW (Swisher, 1999). However, for several reasons, much of the profit from this new capacity is going to overseas manufacturers. First, European wind technology development is subsidized by governments at five times the level of the subsidies in the United States. Second, wind power markets in Europe have been largely inaccessible to U.S. developers and vendors. Third, and perhaps most important, the efforts of U.S. companies to pursue ventures in developing countries have been undercut by European public-private ventures supported by aggressive export promotion programs. These tactics have been very effective at keeping U.S. suppliers out of these markets.

As a result, the U.S. dominance in the development of wind technology in the 1980s, when there were more than 40 developers installing technology in the United States alone, has evaporated. Today, only one major U.S. developer is installing technology, and most wind turbines installed in the United States are imported. Two reasons for the dominance of European manufacturers are: (1) foreign wind turbines have a proven track record in their domestic markets; and (2) they are subsidized by European governments.

Nevertheless, the U.S. market has been making a comeback in the last year. The committee believes that wind energy in this country will best be able to reach its deployment goals by a concentration on larger machines designed for wider deployment. This is one of the lessons of the success of European industry, that, with the right incentives, could be applied in the United States. Smaller machines may be appropriate for niche or remote markets, but the committee believes they address a lower priority market demand than larger machines.

Findings and Recommendations

Finding. Wind power technology is one of the most mature renewable energy technologies, and most pressing technical issues are related to near-term commercialization.

Recommendation. Better coordination between the Wind Energy Program and state programs will be essential for maximizing the efficiency of overall wind power development. The Wind Energy Program should renew its focus on resource assessment, especially on the development of assessment tools for characterizing local wind resources and incorporating wind generation into utility system planning and forecasting models. Storage and other aspects of power system integration should have a high priority in the strategic program plans.

Finding. The domestic wind technology industry is in decline. Most current installations are using imported equipment because heavy subsidies by foreign governments have undercut U.S. competitiveness. As a consequence, U.S. industry is not likely to maintain the technological lead in advanced wind power technology.

Recommendation. Research by the Wind Energy Program on advanced wind turbine technology should focus on turbulent flow studies, durable materials to extend turbine life, blade efficiency, and higher efficiency operation in lower quality wind regimes. Research in these areas should be supported by the Wind Energy Program. The development of advanced controls and improved gearboxes appears to be well within the capabilities of industry.

Finding. The recent Wind Powering America initiative is not coordinated well with the current activities of the Office of Power Technologies' Wind Energy Program.

Recommendation. The U.S. Department of Energy should coordinate the activities and goals of its Wind Energy Program with the Wind Powering America initiative.

Finding. Wind energy for power production is a global market.

Recommendation. The U.S. Department of Energy should investigate the potential of the global wind energy market. Overseas markets may be essential for a struggling U.S. industry. Special requirements in these markets may include technology requirements, such as power system integration and a demonstrated ability to operate under very different environmental conditions.

Finding. Most environmental concerns associated with wind power technology have been addressed.

Recommendation. The Wind Energy Program should take into account outstanding environmental concerns associated with wind turbine siting decisions and in the development of next-generation wind turbine designs.

CROSSCUTTING PROGRAMS

OPT is home to several programs that are not in themselves designed to generate electric power. These include programs on transmission reliability, superconductivity, energy storage, distributed power generation, electric industry restructuring, international markets, and resource assessment. OPT programs generally focus on the real-world use of the technologies under development in other OPT programs.

The real world, where technologies are integrated (as opposed to merely demonstrated) into the broader energy economy, runs on crosscutting issues, such as the basic rules of utility restructuring, the political climate, the availability of renewable resources of all kinds, the nature of the transmission and distribution grid (and the rules they operate by), the particular politics of a foreign country, and so on. In the real world, these crosscutting issues are often the most important ones.

Crosscutting issues can and should have a profound impact on the structure of the upstream technology program. For example, if the majority of states have net metering laws, the availability and reliability of complete, modular solar systems will have a significant impact on the market for rooftop solar systems. Thus, balance-of-system technologies and component integration should have high priorities in OPT programs. When the Asian financial crisis occurred in 1998, the outlook for the geothermal power industry was significantly affected. Assessing its effects and responding to the changes it caused required assessments of global resources and the identification of alternative markets. Wind, geothermal, and biomass power are all affected by transmission issues and the possible impacts of FERC rules, especially rules for independent system operations.

All intermittent renewable energy sources (e.g., wind, photovoltaics) have a stake in energy storage technologies. As renewable energy technologies mature toward true market viability, these and other crosscutting issues should be included in the structure of OPT's technology programs. A strong commitment to the integration of renewable energy technologies into the broader energy economy through crosscutting, or matrix, functions could improve the real-world success of all renewable energy technologies.

RESTRUCTURING OF THE ELECTRIC UTILITY INDUSTRY

The era of the traditional, vertically integrated utility as the primary supplier of electrical power may be coming to a close. The framework for providing electricity is changing across the country at the state and federal levels. The FERC (Federal Energy Regulatory Commission) now requires open access to transmission systems, and many states are offering some level of consumer choice. In the states that have taken restructuring the furthest (e.g., California, the New England States), electric utilities have been required to sell off their power-generating capacity, cede control of the transmission system to an independent system operator, and compete with other energy service providers for customers. These changes will have far-reaching implications for renewable technologies and on how OPT fulfills its mission. OPT must continue to analyze public policies, assess these changes, and plan its programs accordingly so that renewable energy sources remain viable real-world options. Generally, the changes caused by restructuring can be grouped into two categories: changes in the market rules and changes in the players.

Changes in Market Rules

Market rules relevant to renewable energy technologies include the rules governing access to the wholesale and spot market for power generators, regulations governing the behavior of utilities, and guidelines for consumer choice. In other words, every stage, from generation of power through the delivery of power to customers could either promote or impede the development of renewable energy sources.

Some changes in market rules are particularly important to OPT's mission. For instance, many spot markets are developing special rules for intermittent sources of energy. Because many renewable energy sources rely on weather-related energy flows for their primary fuel, these rules have the potential to completely stall the market.

But not all rule changes threaten the market for renewable energy. First, a number of states now require that energy service providers maintain a minimum level of renewable energies in the portfolio of energy they buy for their customers. These policies, known as "renewable portfolio standards," may eventually stimulate significant development in the renewable energy industry. Second, green marketing and disclosure, two customer-choice policies, may also stimulate the development of the renewable energy market. Green marketing refers to energy service products that explicitly include energy from environmentally preferable sources. These products can be provided either in a competitive market or through a regulated utility. Disclosure entails providing customers with information about the environmental characteristics of their energy eventually through something like a nutrition label. Finally, DOE, which has traditionally played an important role in helping state regulators determine the implications of various forms of regulation, may be able to help states design regulations that will promote renewable energy technologies.

Striking a balance between mandating support for renewable energies and developing market mechanisms to support them is a process that must be left largely to each state. However, because of OPT's experience in the development of renewable energy technology, OPT can assist states in determining the impacts of their policies and the design of renewable energy programs. OPT's traditional roles of encouraging information sharing, providing policy analyses, and performing program comparisons should be continued.

Changes in Market Participants

Not surprisingly, the changes in the market rules have led to changes in the key participants in various stages of the market. The most obvious change is reflected in the change in OPT's name from the Office of Utility Technologies to the Office of Power Technologies. Regulated utilities are no longer the main developers of power plants. Indeed, a growing number of privately developed "merchant" plants do not have contracts with utilities to buy power. Thus, OPT must take into account a new class of developers and new hurdles for renewable energy technologies (e.g., merchant financing).

Another challenge facing OPT as a result of restructuring is that some companies will be unregulated and competing for business while others will operate in completely or partially regulated monopolies. OPT's support for these camps

TABLE 3-2 State Funding for Renewable Energy Development and Deployment

State	Total Funding 1998–2010 (in $ millions)
California	540
Connecticut	275
Illinois	50
Massachusetts	332
Montana	10
New Jersey	258
New Mexico	40
New York	15
Pennsylvania	31
Rhode Island	10
Total	1,561

Source: Wiser et al., 1999.

will largely determine whether renewable energy sources are developed primarily as an element of the competitive market or as a public policy. As the country moves toward market-based governance, competitive markets will be used to implement public policy. DOE must, therefore, consider how markets can be leveraged to achieve the goals of the CNES.

States are one group of players that is taking on a more direct role in the development of a renewable energies industry. Renewable portfolio standards and/or SBCs (system benefits charges) are part of the restructuring of the electric industry in 13 states. SBC funds, which are intended to continue funding public benefits programs, generally include improvements in efficiency and renewable energy as part of the mix.

Thus, the renewable energy technology community faces a new challenge. The infusion of almost $1.6 billion through 2010 into technology development and deployment (see Table 3-2) is an opportunity that will probably not be continued unless significant renewable energy facilities have been installed by that date.

DOE is in a position to work aggressively with the administrators of these funds to develop program design, coordination, and evaluation. Coordinated technology development and commercialization will enhance the effects of OPT's programs and benefit U.S. companies. In fact, OPT programs will probably not reach their MW capacity goal unless they work with state programs.

Changes in Technology Requirements

The substitution of bulk power markets for centrally planned, regulated utility supplies in many regions of the country raises questions for research about the design and operation of electricity markets, the coordination and dispatching of supplies, and increases in the carrying capacity of existing transmission corridors. The restructured institutions responsible for the supply, coordination, and governmental oversight of the new market have eliminated traditional sources of, and responsibilities for, R&D support. At the same time, the need for better analytic techniques, both to oversee market performance and to enhance systems operation, has increased substantially as a result of the creative, frequently unpredictable, and novel behavior of new competitors (DOE, 1999h).

Overseeing a market for a unique commodity, conveying electricity through a complex network, and ensuring system reliability are public goods that require publicly supported R&D. The newly organized independent system operators (ISOs) across the country are responsible for ensuring open access to transmission facilities for electric power generators, and FERC is responsible for commercial oversight; but the private sector now has few incentives to support R&D on renewable energy technologies. Therefore, R&D can and should be supported by DOE.

In addition to improving the reliability and efficiency of the power system, increasing the capacity and improving the operation of existing electric transmission facilities would also yield benefits for some components of OPT's R&D portfolio. For example, geothermal, CSP, wind farms, and hydroelectric power generating sources are all location specific and usually far removed from the centers of electricity demand. Thus, a robust, well functioning transmission grid will be essential for the successful implementation of these technologies.

Potential opportunities for R&D include: (1) the development of new operating practices with flexible control for transmission systems (FACTS) devices; (2) the development of advanced computer simulations of operating consequences; (3) the development of high-speed, remote sensing and communication of component conditions; and (4) the development of real-time simulations of emerging conditions and preferred operating responses. Integrating these advances into simulations of new market structures, including practices in which market participants may make decisions that may not be best for the transmission system (usually referred to as suboptimal decisions), would facilitate trial and error experimentation through simulation rather than through risky, on-line "experiments," which are likely to slow innovation.

The transmission and distribution system is a network of connected systems, the transmission system delivering bulk power at high voltages to various regions of the country and the distribution system, after suitable voltage reductions, carrying the electricity to a wide variety of commercial, industrial, and residential end users. Some distributed technologies are likely to be implemented at the very end of the line (i.e., at the local level of the electric system). At this level, institutional, technological, and operating uncertainties about the interconnection of distributed technologies will have to be resolved.

ASSESSMENTS OF INDIVIDUAL PROGRAMS 83

Findings and Recommendations

Finding. Compared to the wider understanding of the electrical operating characteristics and dynamic behavior of the bulk electricity transmission network, the understanding of the operating characteristics and behavior of the electric distribution system is minimal.

Recommendation. The U.S. Department of Energy (DOE) should evaluate the effects of restructuring on the U.S. electric distribution system. Because no other single institution has adequate incentives to undertake this evaluation, DOE should support research on the system behavior, operation, and control of the electric distribution system. The successful implementation of distributed power-generating technologies (which is the essence of the Office of Power Technologies' programs) will depend on a proper evaluation and widespread understanding of the evolving electricity distribution system.

Finding. Restructuring will challenge the Office of Power Technologies (OPT) to respond in a logical, coordinated way to market developments. If each OPT program is left to respond to restructuring on its own, precious resources will be wasted in duplicative efforts that will not satisfy overall market needs.

Recommendation. The Office of Power Technologies must maintain its policy analysis capabilities and coordinate resources from its various technology programs to respond to changes in the energy sector.

Finding. In the restructured wholesale market for electricity, electric power will be treated as a commodity. The existing electric transmission networks were not designed for such a system.

Recommendation. The U.S. Department of Energy should conduct research to ensure the reliability and efficiency of the electric transmission networks in a restructured electricity market.

DISTRIBUTED ENERGY RESOURCES

Many of the technologies supported by DOE programs, including solar, small wind, storage, fuel cells, and advanced natural gas turbine systems will enter the market by interconnecting to the distribution system. Technologies supported by DOE programs in the last decade, which are now entering commercial markets, are encountering a host of technical and commercial barriers. In the developing market, the value of distributed resources goes beyond the price of electricity and includes reliability, power quality, combined heat and power, environmental quality, and other factors.

Barriers to new technologies include a wide range of difficulties posed by market shifts to smaller economies of scale. Interconnection is a larger percentage of project costs for smaller emerging technologies than for independent power projects with larger, more established technologies. In some cases, the same transmission-scale requirements for interconnecting large sources are required for new smaller technologies. The reordering of the grid, as well as a need for new rate-making practices to accommodate new distributed-generation technologies, will create significant regulatory and business challenges.

Like the introduction of customer-owned telephone equipment in the 1970s, the transition to a distributed power system will require both technical engineering protocols for interconnection and new regulatory and commercial practices to open the market to new technologies. The grid of the future should be able to accommodate the entry of innovative distributed energy technologies. Research in support of long-term, reliable grid operation, including the effect of interconnecting distributed energy sources to the transmission and distribution system, should be included in DOE's responsibilities.

Distributed power technologies now entering the market face significant barriers to interconnection at any commercial scale. DOE is already addressing the following issues:

- impact of large market penetration by distributed power systems on the power distribution system
- the absence of standards for interconnection
- the absence of national building and safety codes
- institutional and regulatory barriers
- changes in power distribution system technology and operations to enable the benefits of distributed power to be realized

In FY99, OPT undertook the following activities related to interconnection:

- strategic research; initial five-year strategic planning; development of a distributed power road map; estimates of environmental and economic benefits of distributed power
- R&D on system integration; development of national interconnection standards through the Institute of Electrical and Electronics Engineers (IEEE) Standards Coordinating Committee 21 (SCC21) P1547 Distributed Resources and Electric Power Systems Interconnection Working Group
- study of interconnection barriers; a plan for participating in state regulatory processes; convening of a workshop

OPT is considering the following future activities:

- two or three concept studies for operating the distribution system with

distributed power; development of a technology R&D road map for distributed power
- R&D on system integration; continued development of national interconnection standards through IEEE SCC21 P1547; hardware tests to verify performance of interconnection standards; modeling and analysis for distributed power system integration tests to identify safety, power quality, interconnection, and environmental issues related to the widespread deployment of distributed generation and storage; investigation of practicality and value of modular power system interface units to provide compatibility and interconnection for distributed power (referred to as "plug-and-play" compatibility)
- support for the development of model ordinances and national building and safety codes for distributed power; analysis of the impact of policies and regulations on the growth of the competitive market for distributed power; development of methodologies and tools to facilitate stakeholder decisions on distributed power; providing technical assistance to states and other government agencies

In response to changes in the electricity industry, OPT is reorganizing its own structure and has established the Distributed Energy Resources Task Group, under the Power Delivery Program, to focus on distributed power issues.

Discussion

One of DOE's most important activities related to distributed power is working with industry to develop uniform interconnection standards. Working through a collaborative industry organization coordinated by DOE through the National Renewable Energy Laboratory, the IEEE established a Standards Coordinating Committee to develop consensus standards for the interconnection of distributed technologies, including energy storage technologies. Ordinarily, developing standards takes at least five years. Under DOE's leadership, this industry-supported group is working with a two-year time frame. DOE is expected to provide resources, leadership, and technical support.

The same kind of process will be necessary to establish national standards for the next tier of institutional issues, which range from the prohibition of interconnecting distributed technologies under state and local codes to a patchwork of permitting, tariff, and contract practices that are impeding the emergence of new smaller-scale technologies, despite the growing market demand for them. Other institutional issues are effective, reliable operation of the power grid with distributed power technologies, as well as protocols for grid operations and public access to information. National standards and coordinated approaches to these institutional issues will be necessary to ensure that the competitive market will be open to new distributed power technologies.

Institutional issues related to distributed power cover a range of technologies, from improvements in energy efficiency to renewable energy sources to fossil fuels and combined heat and power generation. The effects of the industry trend towards distributed power on renewable energy technologies and on overall energy policy are not clear. For instance, if expanded distributed power markets advance efficient and renewable technologies, the result will be a cleaner energy system. If cleaner technologies lose ground to diesel-powered or other fossil-fuel competitors, the environmental effects will be very different. Local building codes and environmental permitting can determine the market outcome. However, current policy debates on these issues are not informed by systematic analyses of the impact of market penetration by each technology. DOE has a vested interest in the outcome of these debates as they relate to DOE technologies. The national interest, however, spans DOE programs, other government and private programs, FERC regulations, and Environmental Protection Agency regulations.

Findings and Recommendations

Finding. Successful commercialization of many of the technologies under development in the Office of Power Technologies will require that national markets operate under national standards for grid interconnection and reliability that can accommodate these new technologies. However, the U.S. Department of Energy has not yet addressed initial technical questions about optimal grid configuration and engineering for multiple supply sources.

Recommendation. The U.S. Department of Energy should undertake research and development to determine standards for uniform operating protocols, permitting standards, and regulatory requirements that will be conducive to the development of national markets for renewable energy technologies.

Finding. The technologies under development in the Office of Power Technologies face competition in the marketplace from diesel and other fossil-fuel powered technologies for backup energy or specific energy supplies in distributed resource applications. Neither the environmental impacts of these scenarios nor standards for estimating the impact of market penetration of various technologies has been analyzed.

Recommendation. The Office of Power Technologies, in cooperation with the Environmental Protection Agency, should analyze the impact of diesel and other fossil-fuel powered technologies for backup or specific energy supplies and the impact of renewable-energy distributed resources.

Finding. The Office of Power Technologies has no strategic direction or plan for distributed resources.

Recommendation. The Office of Power Technologies (OPT) should develop a technology road map for distributed technologies to define the role of, and program goals for, distributed power systems in restructured electricity markets. OPT could then define the potential benefits of expanded markets for distributed power technologies and an analysis for policy decisions on distributed power markets.

Finding. The requirements for, and impacts of, the widespread adoption of distributed resources on distribution system networks have not been established.

Recommendation. The Office of Power Technologies (OPT) should continue to support the development, testing, certification, and adoption of interconnection standards and interfaces to support the safe, reliable, economic grid connection of distributed power technologies that can be used in all markets in the United States. OPT's current efforts to develop uniform technical interconnection standards should be expanded to include the modeling, verification, and implementation phases of the technical interconnection programs.

Recommendation. The U.S. Department of Energy should support the development of operating parameters, monitoring, and information systems necessary to operate a distribution system with distributed power generation interconnected at multiple locations. The Office of Power Technologies should work with industry to develop monitoring and information systems for reliable operation of the national distribution grid with distributed power technologies.

Findings. Traditionally, standards for the interconnection of distributed generation resources with distribution grids were established by electric utility companies based on locally defined requirements and regulations.

Recommendation. The U.S. Department of Energy (DOE) should facilitate the development of commercial, institutional, and regulatory standards for the purpose of opening national markets to distributed power technologies. Local variations in standards, ranging from building codes and environmental permits to utility practices and tariffs, require national coordination for national markets. DOE should expand its efforts to develop technical interconnection standards and national energy strategies that address institutional and operating barriers to new technologies.

Finding. Many programs in the Department of Energy besides the Office of Power Technologies (as well as other government programs, the Environmental Protection Agency, and the Federal Energy Regulatory Commission) have an interest in distributed resources for electric power generation.

Recommendation. The U.S. Department of Energy (DOE) should coordinate its assessment of the market for distributed power with all relevant DOE programs, other government programs, the Federal Energy Regulatory Commission, and the Environmental Protection Agency.

INTERNATIONAL ISSUES

One reason emerging technologies produced by U.S. businesses have not been widely adopted internationally is the lack of public/private strategies to assist them, including interagency partnerships with business to facilitate entry into international markets, establish interpersonal public/private networks in other countries, and demonstrate the long-term commitment of U.S. business interests. Although deployment is one of DOE's goals, it is not a funded mandate. The international market will offer a substantial opportunity for renewable energy technologies in the next few decades, especially in countries with higher electricity prices than the United States and regions that do not have transmission grids.

Most foreign competitors have established large public/private teams to encourage international business, and the lack of these prolonged cooperative efforts has created enormous difficulties for U.S. business abroad, particularly in the area of energy supply technologies, which are considered quasi-public activities by most developing countries, even in restructured economic environments. Competing foreign industrial interests routinely send teams to secure business in developing countries that include private sector manufacturers, architect-engineer designers, consultants and bankers, as well as representatives of that nation's export-import bank, government officials who may offer grants-in-aid to develop supporting infrastructure, and government R&D experts to provide education and training for the successful operation of the technology. By comparison, U.S. businessmen are frequently solo agents. American entrepreneurs are also often hampered by a domestic political mythology that divides all activities into public and private categories, whereas, in the international arena, public/private partnerships are the rule for competing effectively in the energy supply arena.

U.S. public/private teams must be consistently staffed with capable decision-makers who can establish long-term relationships with public officials and private customers. Those relationships are essential to ensuring that the target nation consistently adheres to established codes and regulatory standards and sets nondiscriminatory prices and practices for U.S. vendors. Equally important, the committee believes OPT could play an important role in providing the technical backup and training for the cost-effective operation and maintenance of U.S. technologies abroad. The availability of technological support would not only demonstrate the long-term commitment of the United States and its businesses, but would also establish the interpersonal networks and trust to facilitate the continued penetration of U.S. businesses. These ongoing relationships would ease the formation of future relationships and would encourage more U.S. business activity in emerging markets.

Participation of the OPT research community in international technology teams would offer researchers the satisfaction of witnessing the successful outcome of the R&D process (the use of emerging technologies) and would sensitize team members to the goals and needs of developing regions. As a result, research priorities might be modified to increase the probability of U.S. technologies being used in emerging markets worldwide. These systems management strategies, which would greatly enhance U.S. international business penetration, would also be helpful to OPT for the systematic development of coherent R&D strategies and for identifying and developing technologies for future domestic and international markets.

However, it must also be recognized that the development of international markets for renewable energy technologies will require a stronger domestic market base. Foreign companies have made inroads into international markets based largely on the strength of their own national markets. Thus, deployment and potential gains for the domestic industry should be coordinated with a strategy for meeting international needs. If renewable energy technologies truly merit a place in the domestic energy future, then the development of a strong domestic market must be a priority.

REFERENCES

Beyea, J. 1999. Biopower and the Environment. Presentation by J. Beyea, consultant and chief scientist (retired), National Audubon Society, to the Committee for the Programmatic Review of DOE Office of Power Technologies, National Academy of Sciences, Washington, D.C., May 10, 1999.

Birkmire, R. 1999. Thin-Film Photovoltaics: Potential and Critical Issues. Presentation by R. Berkmire, University of Delaware, to the Committee on Programmatic Review of the DOE's Office of Power Technologies, National Renewable Energy Laboratory, Golden, Colorado, June 9, 1999.

Brookshier, P., and J. Flynn. 1999. Hydropower Program Briefing. Presentation by P. Brookshier and J. Flynn, U.S. Department of Energy, to the Committee on Programmatic Review of the DOE's Office of Power Technologies, National Research Council, Washington, D.C., July 22, 1999.

DOE (U.S. Department of Energy). 1997. Scenarios of U.S. Carbon Reductions: Potential Impacts of Energy Technologies by 2010 and Beyond. Washington, D.C.: U.S. Department of Energy, Office of Energy Efficiency and Renewable Energy.

DOE. 1998a. Comprehensive National Energy Strategy: National Energy Policy Plan. DOE/S-0124. Washington, D.C.: U.S. Department of Energy.

DOE. 1998b. Strategic Plan for the DOE Hydrogen Program. January 1998. Washington, D.C.: U.S. Department of Energy.

DOE. 1999a. The Office of Power Technologies: About OPT. Available on line at http://www.eren.doe.gov/power/about.html#mission

DOE. 1999b. Supporting Analysis for the Comprehensive Electricity Competition Act. May 1999. DOE COE/PO-0059. Washington, D.C.: U.S. Department of Energy.

DOE.1999c. Office of Power Technologies Hydrogen Research Program. Available on line at http://www.eren.doe.gov/power/hydrogen.html

DOE. 1999d. Office of Power Technologies Concentrating Solar Power Program. Available on line at *http//:www.eren.doe.gov/csp/csp_tech.html*
DOE. 1999e. Office of Power Technologies Concentrating Solar Power Program: Paths to the Future, Five Year Program Plan 1998–2003. Available on line at: *http://www.eren.doe.gov/sunlab/documents/5yr_plan/5yrp_toc.htm*
DOE. 1999f. Office of Power Technologies Solar Photovoltaics Program: About Our Program. Available on line at: *http://www.eren.doe.gov/pv/pvmenu.cgi?site=pv&idx=3&body=program.html*
DOE. 1999g. Wind Powering America, Draft Action Plan. June 18, 1999. Washington, D.C.: U.S. Department of Energy.
DOE. 1999h. Energy Resources R&D Portfolio Analysis: Panel Report to the Research & Development Council, August 1999. Washington, D.C.: U.S. Department of Energy.
EIA (Energy Information Administration). 1999a. Annual Energy Outlook 1999. Washington, D.C.: U.S. Department of Energy, Energy Information Administration.
EIA. 1999b. Analysis of the Climate Change Technology Initiative, April 1999. Washington, D.C.: U.S. Department of Energy, Energy Information Administration.
EPRI (Electric Power Research Institute). 1999. Electricity Technology Roadmap. Vol. 2. Electricity Supply. Palo Alto, Calif.: Electric Power Research Institute.
Fisher, R.K. Jr. 1999. Industry Perspective on Hydro R&D: Current Programs, Needs, and Opportunities. Presentation by R.K. Fisher, Voith Hydro Company, to the Committee on Programmatic Review of the DOE's Office of Power Technologies, National Research Council, Washington, D.C., July 22, 1999.
Gonzales, M. 1999. Statement before the United States House Committee on Science Subcommittee on Energy and Environment, October 28, 1999, by Miley I. Gonzales, Under Secretary for Research, Education and Economics. Washington, D.C.: U.S. Department of Agriculture.
HTAP (Hydrogen Technical Advisory Panel). 1998. Analysis of the Effectiveness of the DOE Hydrogen Program: A Report to Congress as Required by the Hydrogen Future Act (P.L. 101-566). Final Draft, August 1998. Washington, D.C.: U.S. Government Printing Office.
Iansiti, M., and J. West. 1997. Technology integration: turning great research into great products. Harvard Business Review 75(3): 69–79.
Mitchnick, A. 1999. Hydropower Licensing 1999. Presentation by A. Mitchnick, Federal Energy Regulatory Commission, to the Committee on Programmatic Review of the DOE's Office of Power Technologies, National Research Council, Washington, D.C., July 22, 1999.
Mock, J.E. 1999. Geothermal Energy: That Other Source of Renewable Energy. Presentation by J.E. Mock, consultant, to the Committee for the Programmatic Review of DOE Office of Power Technologies, National Research Council, Washington, D.C., July 22, 1999.
Mock, J.E., J.W. Tester, and P.M. Wright. 1997. Geothermal energy from the earth: its potential impact as an environmentally sustainable resource. Annual Review of Energy and the Environment 22: 305–356.
Neuhauser, E. 1999. Biopower: A Private Sector Viewpoint. Presentation by E. Neuhauser, Niagara Mohawk, to the Committee for the Programmatic Review of the Office of Power Technologies, National Research Council, Washington, D.C., May 10, 1999.
NRC (National Research Council). 1987. Geothermal Energy Technology: Issues, R&D Needs, and Cooperative Arrangements. Washington, D.C.: National Academy Press.
NRC. 1999. Biobased Industrial Products: Priorities for Research and Commercialization. Washington, D.C.: National Academy Press.
OPSWH (Office of the Press Secretary, The White House). 1999a. Developing and Promoting Biobased Products and Bioenergy, Executive Order, August 12, 1999. Washington, D.C.: Executive Office of the President.
OPSWH. 1999b. Greening the Government through Efficient Energy Management, Executive Order, June 3, 1999. Washington, D.C.: Executive Office of the President.

ASSESSMENTS OF INDIVIDUAL PROGRAMS 91

OPT (Office of Power Technologies). 1999a. OPT letter response to questions from the Committee for the Programmatic Review of the DOE's Office of Power Technologies: Biopower Program, April 28, 1999.

OPT. 1999b. Letter response to questions from the Committee for the Programmatic Review of the DOE's Office of Power Technologies: Concentrating Solar Power Program, June 1, 1999.

OSTP (Office of Science and Technology Policy). 1997. Science and Technology Shaping the Twenty-First Century: A Report to Congress. Washington, D.C.: Executive Office of the President, Office of Science and Technology Policy.

OTA (Office of Technology Assessment). 1995. Renewing Our Energy Future. OTA-ETI-614 Washington, D.C.: U.S. Government Printing Office.

Overend, R. 1999. Response to NAS-NRC Questions. Presentation by R. Overend, National Renewable Energy Laboratory, to the NRC Committee for the Programmatic Review of DOE Office of Power Technologies, National Research Council, Washington, D.C., May 8, 1999.

Padro, C. 1999. Hydrogen Program R&D. Presentation by C. Padro, National Renewable Energy Laboratory, to the Committee for the Programmatic Review of the DOE's Office of Power Technologies, National Research Council, Washington, D.C., May 11, 1999.

PCAST (President's Committee of Advisors on Science and Technology). 1997. Federal Energy Research and Development for the Twenty-First Century. Washington, D.C.: Executive Office of the President.

Peelle, E. 1999. Biomass Stakeholder Views and Concerns: Environmental Groups and Some Trade Organizations. ORNL/TM-1999/271. Oak Ridge, Tenn.: Oak Ridge National Laboratory, Bioenergy Feedstock Development Program.

Reicher, D. 1998. Growing an Industry: Overview of DOE's Bioenergy Activities and Proposed Plan of Activities. Washington, D.C.: U.S. Department of Energy, Office of Energy Efficiency and Renewable Energy.

Reynolds, P., M.S. Hay, and M. Camp. 1999. Global Entrepreneurship Monitor 1999: Executive Report. Kansas City, Mo.: Kauffman Center for Entrepreneurial Leadership at the Ewing Marion Kauffman Foundation.

Rocheleau, R. 1999. Comments on the U.S. DOE Hydrogen Program: The Role of Applied Research, by R. Rocheleau, director of the DOE Center for Excellence, University of Hawaii, to the Committee for the Programmatic Review of the DOE's Office of Power Technologies, National Research Council, Washington, D.C., May 11, 1999.

Swisher, R. 1999. National Research Council DOE Programmatic Review. Presentation by R. Swisher, American Wind Energy Association, to the Committee on Programmatic Review of the Office of Power Technologies, National Research Council, Washington, D.C., July 23, 1999.

Thresher, R.W., and S.M. Hock. 1999. Wind Technology Research and Development: Directions and Challenges. Presentation by R.W. Thresher, National Wind Technology Center, and S.M. Hock, National Renewable Energy Laboratory, to the Committee on Programmatic Review of the Office of Power Technologies, National Research Council, Washington, D.C., July 23, 1999.

Wiser, R., K. Porter, and S. Clemmer. 1999. Emerging Markets for Wind Power: The Role of State Policies under Restructuring. In Proceedings of WINDPOWER 1999. Available on CD-ROM only. American Wind Energy Association, 122 C Street, NW, 4th Floor, Washington, DC 20001.

4

Overall Assessment of the Office of Power Technologies

The previous chapter focused on the individual technology programs in OPT. In this chapter, the committee presents a number of findings and recommendations based on the reviews of OPT programs, presentations to the committee, answers to questions submitted by the committee to OPT, the history of OPT, and the personal judgment and experience of committee members.

GENERAL FINDINGS

The committee is encouraged by the changes that were being implemented during this review. For example, the assistant secretary for energy efficiency and renewable energy (EERE) and the deputy assistant secretary for power technologies were reorganizing OPT to the extent allowable under governmental constraints. The reorganization included the development of a strategic plan, an attempt to capitalize on the synergies between the technology development programs, and the hiring of new people. The committee also commends and encourages the efforts of OPT to develop a constructive relationship with Congress.

Efforts are underway to improve communications between the OPT headquarters staff and management, and the staff of the National Renewable Energy Laboratory (the only national laboratory that reports directly to the Assistant Secretary for Energy Efficiency and Renewable Energy), and the other national laboratories (notably Sandia, Oak Ridge, Pacific Northwest, and Lawrence Berkeley). The national laboratories provide much of the scientific and technical expertise available to OPT, and their expertise (along with the expertise of industry and research universities) could be used by DOE headquarters in the development and execution of technical programs.

Finding. The assistant secretary for energy efficiency and renewable energy and the deputy assistant secretary for power technologies are initiating positive changes in organization, strategic planning, and new staffing.

OPT's fundamental problem is bringing technologies to the deployment stage and making a significant contribution to the U.S. electric energy supply system. For many technologies (e.g., wind, geothermal, and solar power), goals and objectives for cost and technical performance have been met; costs have declined substantially; and our understanding of the advantages and disadvantages of the technologies has improved. Nevertheless, partly because of changes in market conditions for electricity production, the deployment goals for renewable technologies have not been met. DOE argues that more R&D should be done to bring the costs down further, advance the engineering and science of enabling technologies, and to identify research that will enhance the competitiveness of these technologies. Before OPT continues with R&D, a thorough road map of each technology should be developed, along with associated cost analysis models, to show the net present value of the technology and the cost required to make the technology competitive.

Finding. Even though substantial improvements in performance and substantial reductions in cost have been made in the last two decades, DOE's deployment goals have not been met.

A number of factors during the 1990s contributed to poor strategic leadership for the DOE R&D portfolio for renewable energy technologies. Congressional efforts to balance the national budget in the 1990s have constrained discretionary funding for energy R&D. In addition, competing national needs, as well as relatively stable and even declining energy prices and no sense of crisis, have decreased public focus on energy issues. The result has been cutbacks in DOE programs and staff. Fewer new people are being brought in, the DOE workforce is aging, and many technical managers have left leading to a decline in experienced, technical leadership. In addition, available technical experts and advisors have not been used effectively. For example, personnel could have been brought from the national laboratories to fill technical advisory positions at DOE headquarters. Unfortunately, many federal work rules make it difficult for DOE to use experts from outside the department.

Attracting highly qualified technical leadership will be critical for OPT's program. Perhaps a rotation system between national laboratory personnel and federal employees could be established. Alternatively, OPT could consider "borrowing" personnel from universities or industries for project assignments as the Defense Advanced Research Projects Agency and NSF do. Other programs, such as programs by the American Association for the Advancement of Science or the American Physics Society, could also be investigated.

The quality of leadership of the individual OPT programs is uneven. Given the strategic thrusts of individual OPT programs and the whole portfolio of OPT activities, highly qualified personnel will be essential to the planning and analysis determining OPT and program priorities.

Finding. Not enough strategic planning and analysis have been done for renewable energy technologies.

The problem of isolated technology programs ("stovepipes") competing for limited DOE resources has been identified in a number of studies and reports over the years. The committee recognizes the value in having separate technology groups work toward their own goals and the value of competition. However, the committee believes that stronger OPT leadership and the formation of cross-cutting teams could help identify synergies among the programs, which would benefit greatly from coordination, as well as a policy focus, especially in light of the significant changes that are taking place in the electric power industry. Although each program seems to have reasonably well thought out objectives, they have not been considered in the overall context of OPT or in light of the changing needs in the electric power sector. There is no strategic approach to R&D that is uniformly understood across all the OPT programs. A number of integrating themes, such as restructuring in the wholesale power market, storage technology, and international opportunities could be used as a basis for changing the focus and objectives of OPT's technology programs.

Finding. OPT programs have operated as relatively separate units with no coordination or integrated planning.

The ongoing restructuring of the electric power sector in many states is resulting in deregulation and cost competition for electric power generation at the wholesale level. Many utility companies are being forced to divest themselves of their power generation assets. Independent power producers are entering the market, and the former "customer" (i.e., the utility industry) for the technologies under development by OTP is rapidly being replaced by diverse agents building and operating their own facilities for electric power production. Although the new environment is reflected in the office name, the Office of Power Technologies, the programs have not been changed.

Finding. In many regions of the country, the traditional customer, the utility industry, for the technologies under development is rapidly changing. OPT programs have not been revised accordingly.

Restructuring in the electric power industry, and the divestiture by many former utilities of their electric power production and R&D programs, have led the private sector to focus on short-term development with payback periods of less than five years. In this environment, much less attention is being paid to long-term issues and the development of new electric power technologies.

Although no one in the private sector is funding long-term R&D, additional R&D will be necessary to bring these new technologies to the marketplace, which will also require an adequate supply of engineering and scientific researchers. Therefore, the state and federal governments will have to underwrite the continuing development of renewable energy technologies. For example, as a result of deregulation, the California Energy Commission's Public Interest Energy Research Program has funds available for energy R&D for the next four years.

Finding. Restructuring of the electric power industry will mean that more of the R&D on renewable energy technologies will have to be underwritten by states and the federal government.

OPT will have to undertake a detailed analysis of the energy market to determine how well various renewable energy technologies can be used in various applications. The OPT programs have not focused on the attributes of a given technology, or hybrid system of technologies, that would enable them to succeed in the market or on the development of information to answer detailed questions (e.g., with regard to environmental impact statements, interconnection protocols, standards, etc.) that are sure to arise during deployment. Technology costs should be commensurate with the value and use of a given technology in the market.

Finding. OPT must develop a better rationale for matching program goals and resources.

Technology road maps are valuable tools for identifying R&D that can enable a vision for an industry to become a reality. The most effective road maps would be developed by industry and then translated to an R&D agenda for OPT.

At the beginning of the committee's study, OPT had not undertaken a coherent roadmapping exercise that included technical objectives and critical barriers to be overcome. A program for achieving objectives and setting priorities, budget requirements, and contingency plans for coping with uncertain budgets needs to be developed. However, the committee was encouraged and pleased to note that during its review, OPT intended to include a road map exercise as part of its newly initiated strategic plan. Although individual program areas have identified critical barriers to the development of their technologies, no systems analysis framework has been used to evaluate the existing and emerging electric power system in detail and to estimate the contribution (e.g., baseload, intermediate load, peaking, hybrid, etc.) of the renewable energy technologies. A systems analysis would reveal the critical R&D areas that require federal support.

The roadmapping process requires some assessment of the full life cycle of energy systems, including supply, processing, distribution, and end use. Therefore, roadmapping is a good way to involve the private sector in technology development and demonstration; and it also facilitates deployment (or market readiness) of the technology. The private sector is then responsible for commercializing the technology.

Finding. OPT has only recently begun to develop technology road maps.

Criteria can be used, along with a road map, for determining research priorities and the role of the public and private sectors in the development of renewable technologies. Criteria also provide a basis for allocating federal funds consistently and systematically. Government participation will be critical in the electric power sector because long-term R&D supported by the private sector has been virtually eliminated. Well developed criteria can also depoliticize debates about the role of the government in R&D.

Finding. OPT has not developed criteria or a systematic process for determining priorities for federal R&D.

The processes and decision criteria OPT uses to select R&D projects were not defined to the committee. A systematic planning process with a range of options for R&D projects might help OPT make decisions that are more acceptable to stakeholders of given technologies. The participation of stakeholders (including advocates and opponents of a given technology) in the establishment of criteria might provide a realistic perspective on the competition facing various technologies and the requirements for success.

Finding. The selection process for R&D project selection is not well defined.

It is not clear to the committee that OPT has a systematic process for balancing short-term and long-term R&D. However, the committee recognizes that OPT is in a difficult position because some projects are congressionally mandated or created by decisions made at other levels of DOE. Nevertheless, some programs (e.g., R&D on hydrogen or superconductivity) are clearly long range and are unlikely to have any impact in the next few decades. Other technologies are quite close to maturity in terms of technical performance and are either unlikely to have much of a market (e.g., large-scale solar thermal power plants) or an industrial base (e.g., hydroelectric power or hydrothermal geothermal power) that could carry the technology into the marketplace.

Crosscutting R&D (e.g., transmission and distribution, energy storage technologies, or distributed power systems) would serve OPT's needs, as well as the needs of other offices in DOE. Goals for crosscutting programs should befit their importance, and DOE should argue for strong internal DOE support, as well as for congressional support of crosscutting R&D. However, OPT does not appear to have made linkages with other government agencies that are funding relevant R&D.

Finding. OPT has not established a systematic approach to determining the balance between short-term and long-term research and development.

In presentations to the committee, OPT program managers did not address (in any substantive fashion) lessons learned from past failures. This lack of attention may be symptomatic of the general tendency of R&D managers to highlight successes and downplay failures. Nevertheless, identifying the causes of program failures can lead to improvements.

Finding. OPT has not paid sufficient attention to lessons learned from past failures.

OPT is not likely to reach its capacity goals unless it works with state programs. The restructuring of the electric power industry has created new opportunities for the development and deployment of renewable power technologies. In many states, renewable energy portfolio standards and/or funding for public-benefit research and increased efficiency are part of the utility restructuring efforts. Funding to keep renewable energy systems in the power generation mix during the transition to a fully competitive market is available in most public-benefit programs, but this funding will be available for a limited period of time. Thus, the renewables community is facing an opportunity and a challenge.

The infusion of almost $1.6 billion through 2010 for technology development and deployment is an opportunity that will probably not recur. To have the best chance of reaching its deployment goals, OPT will have to work with the state programs. If state programs do not achieve defined goals, it will be difficult to justify continuing the investment at the state level and, perhaps, on the federal level as well. OPT is in a position to work aggressively with the state groups administering public-benefit funds to design appropriate programs. OPT may have to educate the fund administrators, who may have little or no knowledge of DOE programs. OPT will have to be flexible in working with state programs, many of which involve both technology development and commercialization. Coordination with states could also benefit companies and organizations conducting research in DOE programs by increasing the resources available for R&D.

Finding. OPT's activities are not well coordinated with state activities.

Management and staff at OPT and elsewhere in DOE have long been open to international collaboration and the use abroad of OPT-pioneered emerging technologies, particularly in developing countries. However, in many instances, those exchanges arose through the serendipity of scientific curiosity. Today, OPT is making a conscious effort to integrate R&D with the needs and desires of international and domestic markets. Globalization of OPT technologies must include addressing growing international concerns about environmental impacts and the effects of increased trade on local cultures.

As a result of coordination furthered by the President's Office of Science and Technology Policy and the commitment of DOE top management to international activities, the essential linkages with private businesses and public agencies necessary for developing and sustaining an R&D program that can capitalize on international opportunities is being established. By participating in and cooperating with international missions, by assisting in the education and training of potential users and scientists from aboard, and by reappraising R&D strategies in the light of these interactions, OPT management and staff can broaden and tailor their efforts to encourage the long-term acceptance, both here and abroad, of the products of OPT research.

Because an internationally focused deployment strategy involves mostly managerial and interagency activities, it must be strong enough to withstand changes in politics and governments at home and abroad. Many individual, project-based initiatives abroad are already under way. OPT must now institutionalize these efforts so that they occur on a routine basis, and the relationships and outcomes can be tracked and evaluated. Therefore, international activities should be included in OPT's strategic planning process, and outcomes should be benchmarked against the goals of those plans. If educational interactions between OPT professional staff and technically trained officials from abroad become routine, those additional responsibilities should be reflected in OPT's budgets and staffing.

Finding. The international market will offer many opportunities for renewable energy technologies in the next few decades.

OPT programs could be integrated with other DOE programs on the development of integrated systems (e.g., housing). Currently, OPT does not interact much with industry on transmission and distribution issues (a consequence largely of the almost nonexistent DOE budget for transmission and distribution). Nor has OPT developed a mechanism for linking its technology development programs to other R&D programs (e.g., programs in the DOE Office of Science, other DOE engineering research programs, and programs outside DOE).

Finding. OPT should forge stronger links with basic science and engineering research programs in DOE and elsewhere.

All of the technology programs in OPT would benefit from a detailed resource assessment that includes the quality of the resource available on a microscale, rather than on a regional level. These assessments would locate and rate particular sites for quality of opportunity for particular renewable energy technologies. At present, OPT's mapping of available resources is uneven and is funded on a piecemeal basis.

Finding. Resource assessment by OPT could be improved.

OPT could evaluate the effectiveness of policy instruments (i.e., renewable energy portfolio standard requirements, federal tax rebates, home owner tax incentives or rebates for renewable energy systems, and community incentives for small, remote distributed generation) to accelerate the development of renewable energy technologies.

Finding. The eventual deployment of renewable energy technologies may require measures by OPT to stimulate markets.

RECOMMENDATIONS FOR THE OVERALL PROGRAM

Recommendation. The committee encourages and recommends that the Office of Power Technologies (OPT) continue the roadmapping exercise and strategic plan it has initiated. Both the road map and the strategic plan should be consistent with the Comprehensive National Energy Strategy developed by the U.S. Department of Energy. The OPT strategic plan should be developed in collaboration with other agencies and sectors and should be integrated with a society-wide assessment of current activities by government agencies and private industry. The road map should distinguish between (1) those R&D activities that promise to provide collective or public benefits and, therefore, require public oversight and (2) complementary R&D activities that primarily promise private benefits and can be left to the private sector. The roadmapping process should include an evaluation of how the technologies under development by OPT could contribute to the evolving electric power supply system, an identification of barriers to technical and market success, estimates of costs for reaching important milestones, and clarifications of federal priorities for development under budget constraints. Based on the road map, some new programs may be developed, some existing programs may be expanded, and existing programs that do not fit OPT's priorities and guidelines may be eliminated.

Recommendation. The Office of Power Technologies (OPT) should develop criteria, a rationale, and a systematic process for selecting research that should receive federal support in light of private sector and state-level activities. OPT should take advantage of the opportunity created by the restructuring of the electricity market to coordinate its activities with state-level renewable energy programs and assist them in implementing the results of OPT programs and promoting the deployment of OPT-developed technologies.

Recommendation. The Office of Power Technologies (OPT) should develop a robust rationale for its portfolio of renewable energy technology projects that will lead to a sustainable, cost-effective energy supply system for domestic and international markets. OPT in general, as well as individual OPT programs, should de-emphasize optimistic, short-term deployment goals as metrics for defining success. The objectives should be the development of a sound science and engineering base, decreases in cost, improvements in technical performance, and the development of technologies that meet the needs of the marketplace. As technologies approach a level of readiness for the market, deployment strategies should be developed in cooperation with private sector agents, as appropriate, and higher policy levels in the U.S. Department of Energy.

Recommendation. The Office of Power Technologies should develop a systematic process for selecting specific research and development programs. The viewpoints of stakeholders should be considered in the development of selection criteria.

Recommendation. The U.S. Department of Energy should take advantage of existing government policies to promote the use of renewable energy technologies for electric power production by encouraging a public demand for "green power."

Recommendation. The Office of Power Technologies should focus more on integrating its programs, identifying common needs and opportunities for research, and clarifying how the individual programs can further their objectives. Benchmarking and other planning techniques used by industry could be adapted for measuring progress and selecting priorities. The challenges posed by the restructuring of the electric power industry, the use of distributed resource technologies, the need for storage technologies for many intermittent renewable technologies, and opportunities in the international market could be the integrating themes. One mechanism for facilitating integration among the individual programs would be to establish crosscutting teams to identify enabling opportunities and critical roadblocks and/or barriers to the development of technologies.

Recommendation. The Office of Power Technologies (OPT) should consider changing its organization and technology thrusts in several ways. Although the Hydrogen Research Program and work on superconductivity have important ramifications for the long term (and should be supported by the federal government), they should not be evaluated in the same way as emerging energy conversion technologies, such as photovoltaics or biopower. Hydrogen has energy carrier and/or storage capabilities that have long-term potential. OPT should develop a clear strategy for supporting long-term research.

Recommendation. The Office of Power Technologies should develop a clear strategy for the development of mechanical, electrical, or chemical storage technologies. Storage requirements for intermittent technologies should be considered in the context of the overall energy supply system. Today, natural gas turbines and pumped hydroelectric power can be used to provide supplemental energy. But promising "clean" energy carriers for the future (e.g., electricity and hydrogen) will require improved energy storage technologies. A breakthrough in either storage technology could strongly influence the future energy infrastructure.

Recommendation. The U.S. Department of Energy should establish a dedicated office to deal with distributed power systems. Whether or not this office is located in the Office of Power Technologies (OPT), its activities should be integrated with those of OPT.

Recommendation. The U.S. Department of Energy (DOE) should assess the effects of restructuring on the nation's electricity distribution system. DOE should provide support for research on distribution system behavior, operation, and control as a basis for assessing the effects of restructuring on electricity distribution systems. An understanding of these issues will be critical to the implementation of distributed generation technologies (which is the goal of OPT's programs). DOE should investigate the integration of distributed generation technologies into the evolving system. This investigation should be strategically coupled with the OPT program and with related activities in the building, transportation, and industrial sectors.

Recommendation. The U.S. Department of Energy (DOE) should provide funds for the direct support of graduate students through a DOE fellowship program leading to an advanced degree related to renewable energy research and development. This would ensure that an adequate supply of scientific and energy talent is available to the emerging industry and that new and inventive ideas continue to flow into the program.

Recommendation. The Office of Power Technologies (OPT) should institute a process for regular external peer reviews (at least every two years) of its proposed and ongoing projects and programs, as well as its overall goals. As part of the review process, OPT should publicly report how it responds to recommendations from external reviews.

Recommendation. Every Office of Power Technologies program should evaluate its resource assessment needs and should fund them accordingly. Resource assessments should be made in cooperation with the appropriate state agencies.

Appendix A

Biographical Sketches of Committee Members

H.M. (Hub) Hubbard (chair) is retired president and chief executive officer of the Pacific International Center for High Technology Research (PICHTR). Previously, he was the Spark M. Matsunaga Distinguished Fellow in Energy and Environment at the University of Hawaii, chair of the National Research Council (NRC) Energy Engineering Board, and chair of the NRC Board on Energy and Environmental Systems. He has also been director of the Solar Energy Research Institute (SERI) and executive vice president of SERI's parent company, Midwest Research Institute. Dr. Hubbard had an adjunct appointment at the East-West Center and is a former member of the boards of directors of PICHTR, the American Solar Energy Society, the Consortium for Pacific Education, and the Guaranty State Bank and Trust Company (Beloit, Kansas). He has also been a consultant to the Idaho National Engineering Laboratory, Argonne National Laboratory, the Congressional Research Service, and the Secretary of Energy's Advisory Board. Dr. Hubbard's expertise is in management of renewable energy research and development (R&D), technology assessment, and energy policy. He received a Ph.D. in chemistry, with a minor in chemical engineering, from the University of Kansas.

R. Brent Alderfer has just opened a utility consulting practice specializing in distributed and green power markets and regulatory strategies. Most recently, he was a commissioner on the Colorado Public Utilities Commission (PUC) and chair of the Energy Resources and the Environment Committee of the National Association of Regulatory Utility Commissioners (NARUC). In that role, he championed the initiation of several distributed-power-related projects and sponsored the NARUC resolutions supporting open markets for distributed-power technologies. Commissioner Alderfer also chaired the Market Power Resolution Drafting Committee for NARUC and has been a leading spokesman for competitive markets and regulatory innovation in the electricity industry. Before his appointment to the Colorado PUC, Commissioner Alderfer was in private law practice handling commercial and natural resource matters. He is also an electrical engineer and has served as a commissioner on the Colorado Air Quality Control Commission, as an arbitrator and mediator, and as a panel member of the American Arbitration Association. He graduated from Georgetown University Law Center in 1977 and has a B.S. in electrical engineering, with honors, from Northeastern University.

Dan E. Arvizu is group vice president for energy and environment and systems, CH2M HILL. He has been director of the Materials and Process Sciences Center; director of the Advanced Energy Technology and Policy Center; director of the Technology Transfer Center; manager of the Technology Transfer and Industrial Relations Department; supervisor of the Photovoltaic Cell Research Division; supervisor of the Photovoltaic Concentrator Systems Division; and a member of the technical staff for solar programs at the Sandia National Laboratories (SNL). He has also been a member of the technical staff for customer switching systems at Bell Laboratories. He has extensive experience in materials science applications for nuclear weapons and energy systems and the development of renewable energy systems, including solar thermal systems, photovoltaic systems, and concentrating solar collectors. Dr. Arvizu was awarded the 1996 Hispanic Engineers' National Achievement Award for Executive Excellence, and he is a member of several advisory groups, including the Commercialization Advisory Board for the Solar II Central Receiver Pilot Plant. He received his B.S. from New Mexico State University and his M.S. and Ph.D. from Stanford University, all in mechanical engineering.

Everett H. Beckner is the deputy chief executive, Atomic Weapons Establishment, and the former vice president, Technical Operations and Environmental Safety and Health, Lockheed Martin Corporation Energy and Environment Sector. His previous positions include principal deputy assistant secretary, Defense Programs, U.S. Department of Energy (DOE); science advisor to Admiral James Watkins, Secretary of Energy; vice president for energy programs, Sandia National Laboratories (SNL); director of energy programs, SNL; and director of Waste Management Programs, SNL. He is a fellow of the American Physical Society and a former member of the NRC Board on Energy and Environmental Systems. Dr. Beckner has broad experience in a variety of solar energy, fossil energy, advanced nuclear fission, fusion, and waste management technologies, as well as in the management of large R&D programs. He also has experience in defense and defense technology issues, technology transfer programs, and nuclear safety. He has a Ph.D. in physics from Rice University.

Peter D. Blair is executive director of Sigma Xi, a scientific research society. He has held a number of positions related to energy technology, energy policy, and energy economics. At the Congressional Office of Technology Assessment (OTA), he was assistant director and director of the Division of Industry, Commerce and International Security. Formerly, he was program manager of energy and materials. In these positions, he was responsible for OTA's research on energy and materials, transportation, infrastructure, international security and space, industry, and commerce. Dr. Blair was a cofounder and principal of Technecon Consulting Group, Inc., specializing in investment decisions related to, and management of, independent power projects, as well as contract research in the area of energy and environmental systems. His primary areas of interest are energy management, systems engineering, and energy policy analysis. He has a Ph.D. in energy management and policy from the University of Pennsylvania.

Charles H. Goodman is vice president, Research and Environmental Affairs, Southern Company. In this capacity, he is responsible for the customer technologies, power technologies, economic analysis, environmental assessment, and the clean air compliance departments, as well as the Power Systems Development Facility at Wilsonville, Alabama. He has chaired the Environmental Staff Committee of the Business Roundtable and is a member of the Environmental Protection Agency Clean Air Act Advisory Committee a member of the Research Advisory Committee of the Electric Power Research Institute (EPRI), and chairman of the EPRI Environment and Health Business Unit. Dr. Goodman is also involved in a number of activities related to the electric power industry that address the ability of technologies to meet existing and emerging regulatory constraints. He is a spokesman on research, environmental and coal utilization issues for the Southern Company. He has a B.S. from the University of Texas at Arlington and an M.S. and Ph.D. from Tulane University in mechanical engineering.

Nathanael Greene, an energy policy analyst with the Natural Resources Defense Council (NRDC), is actively involved in implementation and coordination to spur the development and adoption of fuel cells and solar photovoltaics in the northeastern United States and has worked with the utility industry in a number of states to implement demand-side management programs. He also worked with the Metropolitan Transit Authority in New York to develop models of air quality and the impact of alternative-fueled vehicles and has held positions and been a consultant with the Pace Energy Project, Lawrence Berkeley National Laboratory, the Energy Foundation, and Brown University. He has a B.A. in public policy from Brown University and an M.S. in energy and resources from the University of California, Berkeley.

Jeffrey M. Peterson, program manager, Energy Resources Group, New York State Energy Research and Development Authority, oversees a diverse research program for renewable (photovoltaics, wind, and biofuels) and fossil energy resource development that includes cooperative initiatives to introduce new energy and environmental technologies into the marketplace. He is also currently working with the Center for Clean Air Policy on a World Bank project to determine the potential role of biomass to meet economic and environmental needs in Hungary. He is a member of the Technical Advisory Board, State University of New York College of Environmental Science and Forestry, Center for Forestry Research and Development and the Steering Committee, U.S. Department of Energy Northeast Regional Biomass Program. He was a member of the Technical Advisory Board, Cornell University Center for Advanced Biotechnology, and the External Review Panel, National Renewable Energy Laboratory Terrestrial Biomass Project. He has extensive experience in biomass energy and the development of other renewable energy technologies. He received a B.S. and M.S. in wood science and technology from the University of Massachusetts, and an M.S. in industrial administration from Union College.

T.W. Fraser Russell (NAE), the Allan P. Colburn Professor of Chemical Engineering at the University of Delaware, has also been chairman and professor in the Department of Chemical Engineering, acting dean and associate dean in the College of Engineering, and director of the Institute of Energy Conversion, all at the University of Delaware. He has also been a design engineer for Union Carbide Canada; a research engineer for the Research Council of Alberta; a chemist at the British American Oil Company; and a consultant to a number of industries, including E.I. Du Pont de Nemours. He has been extensively involved in the engineering development of semiconductor materials for photovoltaic modules, including manufacture and commercial-scale designs. Dr. Russell has received a number of awards, including the Francis Alison Award, the American Institute of Chemical Engineers (AIChE) Award in Chemical Engineering Practice, the AIChE Wilmington Section Thomas H. Chilton Award, and the American Chemical Society Leo Friend Award. He has a B.S. and M.S. from the University of Alberta and a Ph.D. from the University of Delaware in chemical engineering.

Richard E. Schuler, who currently directs the Cornell Institute for Public Affairs, holds a joint faculty appointment as professor of economics in the College of Arts and Sciences and professor of civil and environmental engineering in the College of Engineering. While on leave from Cornell, Dr. Schuler served as commissioner and deputy chairman of the New York State Public Service Commission from 1981 to 1983, where he was instrumental in implementing structural changes in the regulation of utilities. Prior to that, Dr. Schuler was director of the New York State Public Service Commission's Office of Research. Before returning to graduate school, he was senior fuels and energy economist with Battelle Memorial Institute for two years, and from 1959 to 1968 he was an engineer and manager with the Pennsylvania Power and Light Company. He currently serves on the board of directors of the New York State Independent System Operator. Dr. Schuler received his Ph.D. and M.A. in economics from Brown University. He also earned an M.B.A. from Lehigh University and a B.E. in electrical engineering from Yale University.

Jefferson W. Tester is director of the Energy Laboratory, Massachusetts Institute of Technology (MIT) and H.P. Meissner Professor of Chemical Engineering. He has also held the position of director, MIT School of Chemical Engineering Practice, and was a staff member and group leader for the Hot Dry Rock Geothermal Project, Los Alamos National Laboratory. He has been involved in various areas of research on energy production and environmental control technologies and on energy conversion and extraction technologies and has written or co-authored more than 125 papers and eight books. Dr. Tester is involved in a number of research collaborations, including the Alliance for Global Sustainability project on energy options for a greenhouse gas constrained world. In his capacity as director of the MIT Energy Laboratory, he is responsible for oversight of a wide variety of energy-related technology developments and policy-related studies. He has served on numerous advisory committees, including the Energy R&D Panel of the 1997 President's Council of Advisors on Science and Technology, the NRC Committee on Energy Conservation in the Processing of Industrial Materials, and the NRC Committee on Geothermal Energy Technology. He has a B.S. and M.S. from Cornell University and a Ph.D. from MIT, all in chemical engineering.

APPENDIX B

Committee Meetings and Activities

1. **Committee Meeting, March 4–5, 1999, Washington, D.C.**

    Presentations:

    Overview of the U.S. Department of Energy's (DOE's) Office of Power Technologies (OPT)
    Daniel Reicher, Assistant Secretary for Energy Efficiency and Renewable Energy

    DOE-OPT Program Descriptions
    Daniel Adamson, Deputy Assistant Secretary, Office of Power Technologies

    Solar Photovoltaics Program
    Jim Rannels, Director, Office of Photovoltaics and Wind

    Wind Energy Program
    Peter Goldman, Deputy Director, Office of Photovoltaics and Wind

    Biopower Program
    Gary Burch, Director, Office of Concentrating Solar Power, Biomass Power and Hydrogen Technologies

    Concentrating Solar Power and International Programs
    Gary Burch, Director, Office of Concentrating Solar Power, Biomass Power and Hydrogen Technologies

Geothermal Program
Allan Jelacic, Director, Office of Geothermal Technologies

Renewable Technologies—The EPRI Roadmap
Stephen Gehl, EPRI, Director, Strategic Technology and Alliances

The PCAST View on Renewables
Robert Williams, Princeton University

National Laboratories' View of Renewables' Potential
Stanley Bull, National Renewable Energy Laboratory

2. **Committee Meeting, May 10–11, 1999, Washington, D.C.**

Presentations:

Status of Technology
Ralph Overend, National Renewable Energy Laboratory

Utility Perspective
Edward Neuhauser, Niagara Mohawk

Environmental Perspective
Jan Beyea, Consulting in the Public Interest

DOE Response to the Presentations and Discussions
Raymond Costello, Biomass Power Team Leader

Summary of Hydrogen Technical Advisory Panel report
John O'Sullivan, Electric Power Research Institute

Fuel Cells
John O'Sullivan, Electric Power Research Institute

Applied Research
Richard Rocheleau, University of Hawaii

Hydrogen Program R&D
Catherine Gregoire-Padro, National Renewable Energy Laboratory

DOE Response to the Presentations and Discussions
Sig Gronich, Hydrogen Research Program

3. **Committee Meeting, June 9–11, 1999, Golden, Colorado**

   Presentations:

   PVMat Program
   Ed Witt, National Renewable Energy Laboratory

   PV Module Manufacture
   M. Misra, ITN Energy Systems

   National Center for Photovoltaics
   L.L. Kazmerski, National Renewable Energy Laboratory

   Industrial and Government Experiences in Photovoltaics
   A. Catalano, Independent Contractor

   DOE Center of Excellence (Thin-Film Photovoltaics)
   R.W. Birkmire, Institute of Energy Conversion, University of Delaware

   DOE Response to Photovoltaics Presentations and Discussions
   James Rannels, Director, Office of Photovoltaics and Wind

   Solar Thermal Technologies
   Craig Tyner, Sandia National Laboratory, and Tom Williams, National Renewable Energy Laboratory

   Solar Dishes
   Herbert Hayden, Arizona Public Service Company

   Solar-Trough Technologies
   David Kearney, Kearney Associates

   DOE Response to Concentrating Solar Power Presentations and Discussions
   Gary Burch, Director, Office of Concentrating Solar Power, Biomass Power

4. **Committee Meeting, July 22–24, 1999, Washington, D.C.**

   Presentations:

   Geothermal Energy Research: Its Past, Present, and Future
   J. Edward Mock, Independent Consultant

APPENDIX B

Geothermal Systems and Opportunities in the U.S. and the World
Michael Wright, Energy and Geoscience Institute

The U.S. DOE Geothermal Strategic Plan
Allan Jelacic, Director, Office of Geothermal Technologies

Industry Perspectives on Short- and Long-Term R&D Needs and Opportunities
Louis Capuano, ThermoSource, Inc.

Twenty-five Years of R&D on Hot Dry Rock (EGS) Systems: Lessons Learned
James Albright, Los Alamos National Laboratory

Draft Strategic Plan and Discussion of OPT Reorganization
Daniel Adamson, Deputy Assistant Secretary, Office of Power Technologies

The U.S. DOE OPT R&D Program and Needs
Peggy Brookshier, DOE Idaho Falls Office, and John Flynn, DOE Headquarters

Industry Perspective on R&D: Current Programs, Needs, and Opportunities
George Hecker, Alden Research Laboratory, and Dick Fisher, Voith Hydro Company

FERC's Perspective on Hydropower Needs Relative to Relicensing
Alan Mitchnick, Federal Energy Regulatory Commission

R&D Partnership: Government and Industry
Jamie Chapman, OEM Development Corporation

Wind Energy Technology Overview
Bob Thresher, National Wind Technology Center, and Sue Hock, National Renewable Energy Laboratory

General Industry Perspective
Randy Swisher, American Wind Energy Association

Future Systems: Transmission and Distribution
Stephen Gehl and Bernard Ziemianek, Electric Power Research Institute

Transmission and Distribution and Storage: R&D at DOE
Phillip Overholt and Imre Gyuk, Office of Power Technologies

5. **Committee Meeting, September 16–17, 1999, Washington, D.C.**

Presentations:

Panel on Distributed Power Generation
Joe Iannucci, Distributed Utility Associates, and Stephen Gehl, Electric Power Research Institute

Panel on Utility Restructuring
Diane Pirkey, Office of Power Technologies; Karl Rabago, Rocky Mountain Institute; and Dallas Burtraw, Resources for the Future

Panel on International Issues
Sam Baldwin, White House Office of Science and Technology Policy; Judith Siegel, Winrock Foundation; Malcolm Cosgrove-Davies, World Bank; Robert Dixon, International Programs, Office of Energy Efficiency and Renewable Energy; Rodger Taylor, National Renewable Energy Laboratory

APPENDIX C

Summary of Recent Studies

Several studies in recent years have addressed various aspects of energy-related research and development (R&D). The major conclusions of these studies are abstracted in this appendix to summarize current thinking on critical issues facing R&D and the deployment of renewable energy technologies.

THE YERGIN REPORT

A task force was established by the U.S. Department of Energy's (DOE's) Secretary of Energy's Advisory Board in 1994 to review DOE's R&D programs in terms of DOE's strategic goals and policy priorities, as well as national needs. The findings from the *Final Report of the Task Force on Strategic Energy Research & Developments* (DOE, 1995) that are relevant to the present study are listed below.

Key Findings

1. Energy is fundamental to the functioning of industrial societies. Global energy demand, arising mainly from developing economies, is expected to grow by about 40 percent in the next 15 years.

2. Energy R&D, both public and private, has greatly contributed to successes in the past 15 years—on both the supply side and the demand side. R&D also contributes significantly to higher standards of living by creating new products, new processes, new jobs, and new opportunities. The contributions of DOE's R&D have also been significant.

3. The federal government should not fund R&D that the private sector can and should support on its own. Federal support for R&D is most strongly justified when the R&D serves national interests that would not be satisfied by market action alone.

4. "Cost-sharing" with industry leverages federal R&D spending, introduces market relevance into federal R&D decision making, and accelerates the R&D process and transfer of results into the economy and the marketplace.

5. The traditional division between "basic" and "applied" research is breaking down. The complexity of research problems requires interactivity between the two. The traditional paradigm is being replaced by "concurrent R&D."

6. Although DOE's management of its energy R&D programs has improved in some respects over the years, it could be much more efficient and effective and could deliver more value to American taxpayers.

7. Effective public investment in energy R&D requires continuity—including much longer funding commitments than the yearly congressional budget cycles. This will require new, innovative financing mechanisms.

Key Recommendations

1. DOE should benchmark its own R&D management practices against "best practices" in the private sector and elsewhere in the government.

2. DOE should adopt "best practices," insofar as practicable, and seek appropriate changes in legislation where best practices are legally restricted or precluded.

3. DOE should develop an integrated strategic plan and process for energy R&D and use this process to determine funding priorities and manage a diverse energy R&D investment portfolio. The portfolio should include the following elements:

   - a balance of basic research and applied R&D (including industry cofunded demonstrations)
   - near-term and long-term R&D to provide continuing return on investment and to contribute to the health and vitality of domestic energy industries
   - a continuing commitment to supporting energy efficiency and renewable energy

THE FIVE-LABORATORY STUDY

Five DOE national laboratories conducted a study to quantify the potential reductions of carbon emissions in the United States by (at least) the year 2010 from energy-efficient and low-carbon technologies (EERE, 1997). *Scenarios of U.S. Carbon Reductions: Potential Impacts of Energy Technologies by 2010 and Beyond* focused on how different sectors of the economy might respond to programs for reducing carbon emissions. Several options for the (electric) utility sector were assessed, and three conclusions emerged from the analysis:

- A vigorous national commitment to develop and deploy energy-efficient, low-carbon technologies has the potential to restrain growth in U.S. energy consumption and reduce carbon emissions by 2010 to near 1997 levels (for energy) and 1990 levels (for carbon).
- Implementation of suggested carbon-reduction scenarios could yield energy savings roughly equal to or more than cost. Only technologies thought to be cost effective by 2010 were studied. Specific policies, political feasibilities, and pathways to achieve the analyzed scenarios were not included.
- The next generation of energy-efficient, low-carbon technologies could enable an aggressive pace of carbon reduction over the next quarter century.

The study found that renewable energy technologies have great potential for reducing carbon emissions, but mostly beyond the 2010 focus of the study. Renewable energy technologies were considered to be in transition from "advanced technologies" to mainstream "technologies of choice" that could play a market role as the cost of generating electricity from these technologies declines. In the analysis of the impact of renewable energy technologies in 2010, a policy of a $50/metric ton cost on carbon emissions was assumed.

Cofiring with biomass was considered to have the technical potential to replace at least 8 gigawatts (GW) of U.S. coal-based generating capacity by 2010, and perhaps 26 GW by 2020. Demonstrations have shown that few modifications to burners and feed-intake systems would be required to cofire coal with up to 15 percent biomass. A dedicated feedstock supply system for fast-growing sources of biomass, such as willow trees, poplar trees, and switch grasses, would have to be developed for biomass to reach a potential of 8–12 GW by 2010, with a reduction in carbon emissions of 16–24 million tons carbon (MtC).

The most important R&D for biomass are in the areas of gasification/conversion systems and feedstock production. Gasification is a demonstration technology for converting solid biomass material to a gas that can be cleaned and burned in a combustion turbine or used in a combined-cycle plant; gasification could double the efficiency of current biomass power. New biomass species could improve crop yields and lower feedstock costs. The development of whole-tree processing methods would lower handling and processing costs.

Wind power technology has been progressing rapidly since 1980, and 1,800 MW of electricity are now produced in the United States. Costs were projected to decline to below the median price for electricity by 2010, with a range of 5 GW (simple extrapolation of growth) to 50 GW (assuming competitive pricing and policies emphasizing control of carbon emissions) of capacity from Class 5 and 6 sites. In this scenario, carbon reductions could be 6–20 MtC. Grid connectivity on this scale may be a problem because of the intermittency of the load.

Wind turbine design is a critical area for R&D to improve materials, increase efficiencies, and lengthen operating lifetimes. Engineering processes must also be improved. Improvements in turbine blade interfaces, with modeling of interactions, could minimize material utilization and extend blade life. Improved direct-drive generators and power electronics should yield higher power conversion efficiencies, perhaps eliminating the need for a mechanical gearbox in the drive train. Better resource characterization of wind prospecting and prediction could help with locating and siting projects.

Hydropower supplied about 10 percent of electricity and constituted 84 percent of renewable energy generation at the time of the study. Hydroelectric technology for utility-scale operations was considered mature, with progress being made to mitigate adverse environmental effects (but not greenhouse gas emissions), such as fish kills, erosion, and water pollution. Three types of hydropower facilities are in operation: dams with storage reservoirs; run-of-river systems without storage reservoirs; and pumped storage projects. Although pumped storage is not a renewable energy technology, it has the potential to reduce greenhouse gas emissions. Further expansion of hydropower capacity may be limited because of relicensing issues and environmental mitigation regulations. Net additions by 2010 are likely to be in the 10–16 GW range, with the potential to reduce carbon emissions by 3–5 MtC.

Costs for solar photovoltaics are currently significantly higher than for other renewable energy technologies, but sales and applications of systems are growing. Off-grid applications for village power are one important growth area. Another is building-integrated photovoltaics, in which solar panels are incorporated into the exterior surfaces of buildings. Thus, grid power would be displaced at the end point of the delivery system where the value is greatest, and photovoltaics peak power output would generally coincide with peak electricity demand. By 2010, photovoltaic installations may have the capacity to supply from 1–7 GW of electricity (based on incentives) and to reduce carbon emissions by 1–2 MtC. With technological advances, costs are expected to be substantially lower than present costs. More progress could be made in the development of photovoltaic power products and systems, as well as improvements in balance-of-systems components, such as power conditioners and controllers.

Geothermal power-generation technologies that produce electricity directly by thermal energy to a steam turbine or via heat transfer to a working fluid that drives a steam turbine were considered fairly mature. Approximately 3 GW of geothermal capacity is currently installed in the United States, with the potential for another 5 GW by 2010. The major problem is locating and characterizing the size and longevity of geothermal reservoirs. By 2020, improvements in drilling technology, seismic data-gathering techniques, and better computer modeling should make location and assessment of geothermal resources more efficient.

Solar thermal electric technologies use mirrors to concentrate reflected sunlight, thus creating a high-temperature source that can be used with a heat engine to generate electricity. There are three types of solar thermal power systems: parabolic troughs (large fields of reflectors heat a fluid in a receiver pipe located along the focal line of the reflector); solar thermal power towers (mirrors reflect sunlight to a thermal receiver atop a tower); and dish/engines (parabolic mirrors in a dish reflect sunlight onto a Stirling engine at the focal point of the dish). By 2010, up to 2 GW of solar thermal capacity will be operational, reducing carbon emissions by up to 1 MtC.

THE ELEVEN-LABORATORY STUDY

In 1997, 11 U.S. national laboratories completed a study of ways to reduce greenhouse gas emissions without inhibiting economic growth. *Technology Opportunities to Reduce U.S. Greenhouse Gas Emissions* (DOE, 1997), known as the Eleven-Laboratory Study (11-Lab), undertook to answer the following questions:

- Which technologies can be improved through R&D that are not now deployed or used extensively?
- Which new technologies could be developed in the future with reasonable effort and cost?
- What kind of R&D program would bring about these results?

The major findings relevant to renewable energy technologies are summarized below.

Renewable energy pathways using energy from sunlight, wind, rivers, and oceans, heat from the planet, and biomass all have the potential to reduce greenhouse gas emissions by displacing fossil-fueled electricity generation or petroleum transportation fuels. In the power sector alone, renewable energies would be capable of reducing carbon emissions by about 70 MtC per year. The costs of renewable energy technologies are decreasing to the point that commercialization is a real possibility for early in the twenty-first century and are already competitive in certain niche markets.

Biomass as a cofired fuel with coal, gasified to replace natural gas, or as a stand-alone fuel has the potential to reduce fossil-fuel-fired electric power generation. R&D challenges that must be overcome are emissions of nitrogen oxides, ash chemistry, and associated operational problems.

Wind energy systems are competitive (on a levelized cost basis) with current power-generation systems. Most states have sites with high-quality wind resources and, if this technology were fully developed, carbon emissions could be significantly reduced before 2010. The next steps toward increasing market penetration are improving the design and reliability of turbines and methods of improving generation at sites via hybridization with other power technologies or new storage technologies.

Hydropower currently accounts for 10 percent of U.S. power generation, but prospects for further development of hydropower resources are not good. Concerns about impacts on fish and downstream water quality will have to be addressed and the cost-effectiveness of retrofitting demonstrated.

Solar photovoltaic technology using semiconductor-based cells to convert sunlight to electricity can work on a variety of scales. Annual growth of the market is 15 percent to 20 percent. The technology works especially well in off-grid applications, but costs are currently too high for bulk power generation. By 2010, photovoltaics could compete for peak power shaving opportunities (when demand for electrical capacity results in high electricity prices) and by 2020, for daytime electric power opportunities. Much research has yet to be done on materials and processes as a basis for advanced photovoltaic cell design and engineering.

Geothermal energy technologies use thermal energy from the earth to produce electricity or heat for industrial processes. Hydrothermal reservoirs produce about 2,100 megawatts electric (MWe) annually in the United States. Direct use of geothermal energy accounts for 400 megawatts thermal (MWt), and geothermal heat pump systems (using the earth as a heat sink for heating or air conditioning) contribute another 4,000 MWt in energy and are growing at about 25 percent per year. Currently, only a small portion of the huge geothermal resource can be used economically, but further engineering and reservoir research could double the production of electricity.

Solar thermal technologies, which concentrate sunlight to generate electricity, have been successfully demonstrated in nine commercial plants that provide 354 MW of electricity in California. Relatively conventional technology could add hundreds of megawatts of peaking power by 2005; further R&D will be necessary for bulk power generation by 2020.

Meeting the goals described in the 11-Lab report will require both incremental improvements and breakthroughs via basic and applied research. Strategic public/private alliances will be the best approach to developing and deploying most technologies for the reduction of greenhouse gases. Public/private strategic alliances will help maximize innovations by bringing together stakeholders capable of overcoming scientific, technical, and commercial challenges. This report describes the reductions in carbon emissions that could result from an accelerated R&D program but does not describe collateral benefits of complementary deployment programs or policies to stimulate markets for these technologies. The most cost-effective approach would be science and technology combined with deployment programs and supporting policies.

The report concluded that a national investment in a technology R&D program over the next three decades would provide a portfolio of technologies that could significantly reduce emissions of greenhouse gases over the next three decades and beyond. A strategic plan that includes deployment policies to complement R&D will be necessary for success. Plans should reflect the economic and technological implications of deploying these technologies. Hence, the development of a technology strategy for mitigating climate change was the recommended next step. The development process should include a review of technology policy options to complement technology development options and a detailed plan for supporting implementation that addresses technology goals, R&D program plans, policies that support deployment, and fiscal resources. Development of this agenda should be a collaborative effort between government, industry, business, and the scientific communities.

PCAST I

In November 1997, the President's Committee of Advisors on Science and Technology (PCAST) completed a study, *Federal Energy Research and Development Challenges of the Twenty-First Century*, that focused on major challenges for a range of energy technologies (PCAST, 1997). The following discussion summarizes the findings and conclusions related to renewable energy technologies.

The primary challenge facing renewable technologies is relatively high unit costs, but progress on that front is being made. The cost of energy from wind power and photovoltaics has decreased about tenfold. Much of the market growth for renewable energy sources is expected to come from developing countries because the small scale and modularity of these technologies is suited to their needs. The panel concluded that R&D spending for renewable energy should be significantly increased. Suggestions were also laid out for improving the efficiency of wind power and photovoltaic systems, as well as the following time-defined technological goals:

1. For wind systems, reduce the cost of generating electricity by 2005 by 50 percent so as to be competitive with fossil-based power generation in a restructured electricity industry.
2. Pursue R&D in solar photovoltaics to reduce the cost of photovoltaic systems to $3,000/kW in five years; to $1,500/kW in 2010; to $1,000/kW

in 2020. R&D should also focus on balance-of-systems issues and advanced materials.
3. Strengthen R&D for solar thermal technologies, such as parabolic dish and heliostat/central-receiver technology with high-temperature storage. Develop high-temperature receivers combined with gas turbine-based power. The goal is to make solar-only power competitive with fossil-fueled power by 2015.
4. In the next 10 years, commercialize advanced, energy-efficient biopower generation technologies employing gas turbines and fuel cells integrated with biomass gasifiers to exploit the advantages of biomass over coal as a feedstock for gasification.
5. Continue work on hydrothermal systems and reactivate R&D on advanced concepts giving a high priority to high-grade hot dry rock geothermal technology (which has the potential to provide heat and baseload electricity in most areas of the United States).
6. R&D on hydrogen-using and hydrogen-producing technologies should be supported. R&D on hydrogen-using technologies should be coordinated with proton-exchange membrane fuel-cell vehicle development by DOE. Working with DOE's Office of Fossil Energy program, R&D in hydrogen production should be prioritized to optimize the production of hydrogen from fossil fuels and the sequestration of carbon dioxide separated out in the production process.
7. R&D for a new generation of hydropower turbines should focus on turbines that are less damaging to fish and aquatic ecosystems. By deploying these new technologies at existing dams and in new low-head, run-of-river facilities, as much as 50,000 MW could be added by the year 2030.
8. Resource assessment, international programs and analysis, and other crosscutting programs should be strongly supported. Additional R&D should focus on energy storage, electric systems, and systems integration.

Other general recommendations included more coordination and networking across the applied R&D "stovepipes" and with the Office of Science. In fact, one of the suggestions was that up to 5 percent of each applied R&D budget be reserved for collaborative, strategically driven, basic research activities with matching funding from, and supervision by, the Office of Science.

PCAST II

In June 1999, PCAST issued a second report, *Powerful Partnerships: The Federal Role in International Cooperation on Energy Innovation,* which is known as PCAST II (PCAST, 1999). The panel reviewed the U.S. role in international energy innovation and the roles of the public and private sectors in these activities. The panel concluded that energy initiatives have a window of opportunity for attracting private sector capital for energy generation for economic development, as well as for addressing public-good issues globally. The energy technologies and infrastructures developed over the next few decades will have a strong impact on energy costs and end-use efficiencies, greenhouse gas emissions, air pollution, and a range of other factors for most of the next century. The globalization of innovation capacities and tightening constraints on spending for domestic R&D contribute to the attractiveness of international cooperation for developing energy technologies. International cooperation would also enhance the ability of U.S. energy companies to enter some of the largest markets for these new technologies. Energy-related global environmental problems and risks could also be lessened. The panel made the following observations related to energy R&D:

- Accelerated innovation in energy technology can increase the pace and decrease the cost of the adoption of technologies that can improve the health and safety of the environment.
- Innovations in energy are necessary to lower the energy intensity of economic activity, reduce emissions from energy activities, reduce the costs of delivering energy in environmentally sustainable ways, and increase energy options.

The panel cited the following reasons for U.S. participation in international energy projects:

- The pace would be increased and the cost lowered of U.S. acquisition of innovations for domestic use.
- U.S. firms would gain access to large overseas markets for innovative energy technologies.
- The global dimensions of energy challenges would be addressed by accelerated development and deployment of innovations worldwide.

Continued government involvement in energy innovations would serve many needs that transcend private interests (e.g., social, macroeconomic, environmental, and international security concerns). Therefore, the panel recommended that government initiatives be structured to encourage, catalyze, and complement, rather than replace, corresponding activities in the private sector.

Specific opportunities for cooperation identified in the study were initiatives for the development of renewable energy technologies and fossil fuel decarbonization. The panel recommended that a broad-based renewable energy cluster organization be established to accelerate the development and deployment of renewable energy technologies, especially to meet energy needs in rural areas of the developing world. The establishment of a fossil-fuel decarbonization and carbon sequestration cluster was recommended as a multinational collaborative effort to develop technologies that would use fossil fuels economically in ways that resulted in near-zero life-cycle emissions of carbon dioxide. Expansion of the Vision 21 Program at the DOE Office of Fossil Energy was suggested, as well as the development of technologies to make hydrogen from carbonaceous feedstocks and to recover by-product carbon dioxide for safe disposal.

REFERENCES

DOE (U.S. Department of Energy). 1995. Final Report of the Task Force on Strategic Energy Research and Development. Washington, D.C.: Secretary of Energy Advisory Board, U.S. Department of Energy.

DOE. 1997. Technology Opportunities to Reduce U.S. Greenhouse Gas Emissions. Washington, D.C.: U.S. Department of Energy (October). Also available on line at: *http://www.ornl.gov/climate_change*

EERE (Office of Energy Efficiency and Renewable Energy). 1997. Scenarios of U.S. Carbon Reductions: Potential Impacts of Energy Technologies by 2010 and Beyond. Washington, D.C.: U.S. Department of Energy.

PCAST (President's Committee of Advisors on Science and Technology). 1997. Federal Energy Research and Development for the Twenty-First Century. Washington, D.C.: Executive Office of the President.

PCAST. 1999. Powerful Partnerships: The Federal Role in International Cooperation on Energy Innovation. Washington, D.C.: Executive Office of the President.

Acronyms

APS	Arizona Public Service
CNES	Comprehensive Natural Energy Strategy
CSP	concentrating solar power
DOE	U.S. Department of Energy
EERE	Office of Energy Efficiency and Renewable Energy
EGS	enhanced geothermal systems
FERC	Federal Energy Regulatory Commission
FY	fiscal year
GEF	Global Environment Facility
HDR	hot dry rock
HTAP	Hydrogen Technical Advisory Panel
IEEE	Institute for Electrical and Electronics Engineers
ISO	Independent System Operator
NADET	Natural Advanced Drilling and Excavation Technology
NSF	National Science Foundation

OIT	Office of Industrial Technologies
OPT	Office of Power Technologies
OTT	Office of Transportation Technologies
PCAST	President's Committee of Advisors on Science and Technology
PURPA	Public Utilities Regulatory Policies Act of 1978
R&D	research and development
SBC	systems benefits changes
SEGS	solar electric generating station
USDA	U.S. Department of Agriculture